Simon Barnes is the author of many wild volumes, including the bestselling *Bad Birdwatcher* trilogy, *Rewild Yourself*, *On The Marsh*, *The History of the World in 100 Animals* and *The History of the World in 100 Plants*. He is a council member of World Land Trust, trustee of Conservation South Luangwa and patron of Save the Rhino. In 2014, he was awarded the Rothschild Medal for services to conservation. He lives in Norfolk with his family and horses, where he manages several acres for wildlife.

Also by Simon Barnes from Simon & Schuster:

Epic: In Search of the Soul of Sport

Rewild Yourself: 23 Spellbinding Ways
to Make Nature More Visible

On the Marsh: A Year Surrounded by Wildness and Wet

The History of the World in 100 Animals

The History of the World in 100 Plants

THE YEAR OF SITTING DANGEROUSLY

MY GARDEN SAFARI

SIMON BARNES

Illustrations by Cindy Lee Wright

SIMON & SCHUSTER

London · New York · Sydney · Toronto · New Delhi

First published in Great Britain by Simon & Schuster UK Ltd, 2023
This edition published in Great Britain by Simon & Schuster UK Ltd, 2024

Copyright © Simon Barnes, 2023

Illustrations © Cindy Lee Wright, 2023

The right of Simon Barnes to be identified as the author
of this work has been asserted in accordance with the
Copyright, Designs and Patents Act, 1988.

1 3 5 7 9 10 8 6 4 2

Simon & Schuster UK Ltd
1st Floor
222 Gray's Inn Road
London WC1X 8HB

Simon & Schuster: Celebrating 100 Years of Publishing in 2024

www.simonandschuster.co.uk
www.simonandschuster.com.au
www.simonandschuster.co.in

Simon & Schuster Australia, Sydney
Simon & Schuster India, New Delhi

The author and publishers have made all reasonable efforts to contact
copyright-holders for permission, and apologise for any omissions or errors in
the form of credits given. Corrections may be made to future printings.

A CIP catalogue record for this book
is available from the British Library

Paperback ISBN: 978-1-3985-1890-2
eBook ISBN: 978-1-3985-1889-6

Typeset in Palatino by M Rules
Printed and Bound in the UK using 100% Renewable
Electricity at CPI Group (UK) Ltd

To the memory of my father
Edward Barnes
1928–2021
indefatigable dog-walker
lifelong nature lover
and bad birdwatcher
among many other things

*Sitting quietly
doing nothing
spring comes
and grass grows by itself*

                                                                  Basho

*Sometimes I sits and thinks and sometimes
I just sits*

                                                Old saying

Contents

Foreword   1

Departure   5
October   13
November   41
December   64
January   85
February   112
March   137
April   165
May   195
June   225
July   251
August   281
September   308
Arrival   335

Foreword

I used to be important. Taxis, aeroplanes, hotels, press conferences, deadlines, 'something to drink, Mr Barnes', 'copy five minutes before the final whistle, Simon, and your considered pieces half an hour after that'. Movement, travelling; passport, laptop, plug adaptors and toothbrush always in the bag ready to go. Bloody good fun it was, on the whole.

I got less important, but I still travelled a fair bit. And then all of a sudden no one was important, least of all me. I couldn't even travel the dozen miles to Norwich. The time of High Covid, the time of Lockdown, was upon us: terrible, frightening, disturbing, damaging – and yet not entirely short of wonder, at least for those of us not living in a basement flat with an abusive partner.

Stationary, grounded, forbidden to move, I resolved to travel in the only way I could: by staying still. I would sit in the same place, day after day, for a full year and journey through the weeks and months. I would sit in sunshine and storm, blizzard and heatwave. I would become part of the landscape of Norfolk, far less important than the tree above my seat. I would notice things: birds of course, also deer, hares, butterflies, bees, mosquitoes, plants and above all the everyday, every-year miracle of the seasons. I would think, and I would not think. I would use my mind and I would lose my mind, or at least mislay it.

I would sit still: and by sitting still I would travel to places I had never been before, where few people have ever been before. The dangers, such they were, would come from weather, from the testing of my patience, resolve and love, and from the alarming opportunity to let my mind fly wherever it chose, vulnerable to bad thoughts and worse fears. I might even – horror of horrors – be forced to face the fact that, as winter moved to spring and the blackcap raised his voice in song, I was now as unimportant as I always had been.

Before the plane takes off a soothing voice takes the passengers through the housekeeping conventions: seatbelts, overhead lockers, the stowing of tray-tables and the unlikely event of an emergency. So, before we set off on this journey through time, I should make a few announcements.

Most species are given their full common names, which should be self-explanatory. But there are a few exceptions. Unless otherwise stated:

Swan is mute swan

Crow is carrion crow (I think we need the but. Crows are corvids too.)

Corvids usually means a mixed flock of mostly rooks and jackdaws

Kite is red kite

Goose is greylag goose

Harrier is marsh harrier

Egret is little egret

Deer is Chinese water deer, an introduced species common in the Broads

Hare is European brown hare; alpine hares are not a feature of the Broads

I have mostly avoided the abbreviated bird names that birders use but there are a few exceptions:

Cetti is Cetti's warbler

Greatspot is great spotted woodpecker

Kes is, unsurprisingly, kestrel

Pinkfeet are pink-footed geese

When the gender of the observed animal is known, I have used he and she; when not, it. My immediate family who share the place with me are my wife Cindy, who is an artist (www.cindyleewright.com), older son Joseph, twenty-seven who makes and teaches music, and younger son Eddie, who does farming, photography and performance; Eddie is twenty and has Down's syndrome. I hope that's all clear. Now, let's go ...

> Each day I have catalogued is followed by a small chunk of advice and/or information. This is mostly intended for less experienced wildlifers: feel free to skip.

DEPARTURE

Sunday 27th September

I should have been in Africa. I should have been in the Luangwa Valley in Zambia, co-leading the Sacred Combe Safari: walking with lions, elephants round my hut, bateleur eagles in the sky and all the pleasures of sharing and showing and teaching these wonders to our clients. It's a job, of course, but a rather joyful sort of job; one I had taken on several times before – but this was the time of Covid and I was at home. The swifts had gone to Africa, leaving me behind.

So I embarked on the only voyage of discovery open to me. I set off along the path less travelled by, in search of unfamiliar places. I was wearing, among many other items, two fleeces, a waterproof top and waterproof trousers. I felt nervous, excited, uneasy, not sure if I was up to it and whether I would last the pace: as you always do with journeys into the unknown.

I sat on a folding chair at the bottom of my garden. And so it began.

Itinerary: sit here – right here, in this very spot, in this very chair – for the next year. Spend a good part of every day looking at exactly the same view. Some days professional and family obligations would make that impossible, but never, I resolved, the weather. I would sit through the seasons. I would sit here in wind and rain and snow and frost and sun. I would journey from one year's end to the next. I would make a stationary safari.

My point of departure was easy to choose. I have the unearnable good fortune to live in Norfolk, on the edge of the Broads and next-door to a dozen acres of marshy scrub managed for wildlife. Beyond them is a half-mile strip of water-meadow leading to a small hidden river; on each side of which there is a copse of wet-loving trees, technically an alder carr.

It was raining on this, the first day, and that was somehow important. It was like taking my seat at La Scala, except that opera-goers seldom have to tip half a pint of icy water from their seats before sitting down. It was like taking a pew in

Norwich Cathedral below an incalculably lofty ceiling: the vast Norfolk sky vaulted with clouds. Or perhaps it was like entering a Zen meditation hall, to sit still and share the air with willows and alders and sallows.

I could feel the coldness though not the wetness of the water through my waterproof trousers. I could hear the quacking of mallards. Along the line of the river was a line of black-headed gulls. The sky looked like an old grey woolly. Was I here to explore nature? Or my own nature? Even as I posed this question I could feel the foreshadowing of an answer. It took the form of another question.

You mean there's a difference?

Just beyond the river I could make out the silhouette of a marsh harrier, balancing like a tightrope walker on the big wet wind. The year of sitting dangerously had started. The bottomless sit had begun.

> A supermarket plastic bag in your pocket means you always have a dry place to sit.

Monday 28th September

Another day, another sit, another puddle on the seat. I tipped, sat, looked, listened. Beyond the hidden river lies the Flood, also hidden: 100 acres of open water. This floodplain landscape lies at the bottom of what we in Norfolk humorously call a valley, so beyond the river and the Flood the land rises up – gently – to a line of oaks. And above them, above everything, the limitless Norfolk sky.

The alder carr on the left holds in season a small colony

of nesting herons, so I will call it the heronry even when the herons are absent. The one on the right I will prosaically call the right-hand carr. Beyond the heronry – to the north – the sky was lighter, dazzling after yesterday: a shimmering pale grey rather than charcoal. I could hear rumbling caws from the rookery out of sight behind me; a small flock of jackdaws flew across, energetically jacking to each other.

It was still warm enough for insects to fly in quantity over the marsh, living their own lives and feeding the lives of others. A fly orbited my hat and settled there for a moment: I was still enough to count as an inanimate object; an early victory. A rattle of magpie: the day's third crow species. Are they really such grim birds? Vincent's *Wheatfield with Crows* never seems to me a picture of despair.

I sat. Teach us to sit still, said T. S. Eliot in 'Ash Wednesday', sounding more like an exasperated schoolteacher than a master of meditation. *Sit still, Barnes! Barnes, you're a nothing but a flibbertigibbet!* What would I learn in this great classroom where I sat now? A carrion crow gave a harsh triple-caw like a symbolic bird in a problem painting. But that other painting, the old man sitting with his head buried in his hands, that's despair, isn't it? Vincent called it *At Eternity's Gate*.

I sat, not in despair, hearing the sound of a swan's wings. As you sit without agenda, thoughts come into your head without warning: I remembered coming across my mother with her head in her hands. We were in a public garden in Cornwall, a place she loved. She'd had her first stroke by then and walking wearied her. So I walked on a little, leaving her to enjoy a good sit – and when I came she was sitting as if modelling for Vincent. She died all but thirty years ago; my father, now in his nineties, was in good heart, living in London and most days walking as far as the Thames, maybe 300 yards, where he too would have a good sit, watch the river getting on with the job of flowing, binoculars round his neck. When we spoke on the phone that night, he would tell me of any good birds, and how lucky he was to have such a thing as a river so close to home. Lucky us, with our rivers and our birds.

A jay called out, a mad scream: another crow species all right, but coloured in pink and blue. They've been compared to birds-of-paradise, which are quite closely related to crows, as it happens. Five species of crow in a single sit. One day, perhaps, a raven will drop by. A small beetle landed on my leg, shining with iridescence, making my waterproof trousers even more beautiful.

> Making notes concentrates the mind. You don't have to write them up afterwards, or even look at them again: the act of note-taking is significant in itself.

Tuesday 29th September

I tipped the puddle from my seat with a feeling of comfortable routine, and sat. The rain continued; I stayed dry. Good clothing. I listened to the music, mostly percussion. The small drops of rain that hit me directly made a soft pattering. Above my head the ash tree held onto the water for a while before releasing it in bigger drops. These splashed, mostly onto the brim of my hat. It was like being a drum-kit: a fast, light pattern from the snare drum punctuated by steady, powerful beats from the bass.

No matter how good our clothing, rain is something we hurry through. We may stop to smell the roses, but we seldom stop to feel the raindrops, certainly not for a sustained period. This was, perhaps, the point at which my journey of exploration truly began, I thought: exploration of self, exploration of nature. Then I thought that it would be more likely to begin when I stopped congratulating myself.

So what are the rules of the Bottomless Sit? Keep still. Don't get up for anything. Never look at your phone. Make the odd note. Questions of ID and behaviour may only be researched when the day's sit is over. Let nature set the agenda.

Then at last there was melody to add to the percussion. A robin sang out, sweet and soft in the soft rain. I learned to identify birds by song in my late thirties and as a result birdsong is one of my greatest joys: be certain that the voices of birds will echo through this project.

I sat on as the rain fell. I was able to countenance this idleness by calling it work. I've never truly mastered the distinction between work and play. Perhaps that's a bad thing, I don't know; Noël Coward said that work is more fun than fun. This sitting business is actually about not working: but it seems to me acceptable because I am making a work project of it. I hope it works.

Swallows flew overhead, two – no, three of them. Despite the light rain they were finding plenty of insects in the air; I could tell that from the jinks and kinks in their flight, each one a flying canape, a little death. They were getting ready for their journey to Africa. Heigho. For the first time in a good few years I would not be joining them; I would not be pointing out to our guests the songs of orange-breasted bush-shrike and Heuglin's robin: I would be sitting here, seeking new horizons in the familiar jagged line of the trees on the far side of the river.

> 🍃 More or less the only complicated song you will hear in the autumn is the robin's. If you listen out at that time you will get it fixed in your brain. To learn a little birdsong is to go through one of the great portals of discovery: once through it you are an insider. As we move on through this year I will try to help beginners to find the doorway and pass through it.

Wednesday 30th September

The day was grey with a small hint of silver; the chair held a puddle in much the same colours; my mood was similar but without the silver. I felt the chill through my waterproof trousers and resented it. I mean, this is stupid, isn't it? Why don't you give it up as a bad job right now? Where did this bird-brained, hare-brained, altogether lunatic scheme come from? But I didn't even have the heart to walk away. All right, then, let's get this over with; there won't be much out there today, not in this weather; make a nice note about something obvious and get back inside. You can think about giving up properly tomorrow.

That was when a falcon whizzed past over the alders and something in its sharp-winged speed made it clear that this wasn't a kestrel. I had my binoculars on it with pleasing efficiency – and it was a hobby, long-winged, both swift and swift-like, fast enough and agile enough to take swallows out of the sky. Like the swallows, it would be off to Africa soon: it was moving southwards as it vanished.

There's an odd pleasure in a bird that looks like a humdrum species and then reveals itself as something exotic. If you think it might be a hobby, it's always a kestrel – but you seldom mistake a real hobby for a kes. You can mistake the illusion for reality, but you can never mistake the real thing for illusion. Call that Barnes's law and apply it to truth, trust, love . . .

Three mallards flew across the sky as if the tree-line was a mantelpiece.

> 🍃 You can buy waterproof over-trousers for less than ten quid and more than a hundred quid; it's about quality and breathability. Do get hold of some: the point is that any pair will change your world.

OCTOBER

Thursday 1st

Sometimes the joy of finding something is so great it was almost worth the pain of losing it: phone, wallet, best pen, best friend, lover. The reunion reveals a truth: normal life is a fairly blessed state. I tipped the puddle from my seat and sat; the sky was a sable silvered. After a while I heard two notes – two marvellous, lost, half-forgotten notes. And again. By the third time the sound was familiar again. Curlew!

Mostly they come here in the autumn and stay for the winter; in spring they leave to make nests and more curlews on the high tops. Their departure adds a touch of sadness to the joys of spring; their arrival gives delight to the advancing chills of winter. There were just two of them, flying overhead and calling hard; sharp wings, long beaks, white diamonds on their backs.

Cetti's warblers came north to the wet places of southern Britain (mostly below the Wash) on the advancing front of climate change. They stay all year: seldom seen, often heard: strident, comic, assertive, shouty, sweary. They sang out with vigour at every opportunity in the first seven years I knew this place: then came the snow. In February 2018 the Beast from the East arrived and that did for the Cettis. They had missed the last three springs.

I was in the process of deciding that I had done enough sitting for the day when an unmistakable Cetti called from ten yards away: so loud and sudden I almost jumped.

Lost birds. Found birds. From my soggy chair I gave them welcome.

> They're named for an Italian Jesuit naturalist Fr Francesco Cetti, so say *chetty*. If you search the name of a British bird and then add RSPB you will find a picture, a brief description, a guide to where you're likely to see it – and a brief recording of its voice.

Friday 2nd

No puddle on my seat today, a small miracle. It was moving towards dusk; I was late because I had been to London to visit my father. Travel to and across London by train was a worrying business: catching Covid was a bad idea for me, and a worse one for Eddie, what with Down's syndrome and all. I wasn't that keen on introducing the virus into my father's house: but I didn't want to leave him unvisited either.

There was a sparky wind and I was feeling a little turbulent myself. But he had been in good form, certainly not at eternity's gate. He had been delighted when I told him of the return of the Cetti, recalling an occasion we had looked for them in Cornwall, and remembering my mnemonic for their song. Glass of Sauvignon Blanc in hand he had shouted: 'Me? Cetti? If-you-don't-like-it-fuck-off!'

Quite soothing, this turbulent view, with its wind-tossed jackdaws, wood pigeons and black-headed gulls, sky in lavish Norfolk quantities. And then a fragment of robin: say two seconds of sweet song.

Well, it sounded sweet, even though the robin was expressing the same sentiments as the Cetti. Then I remembered the cooing tone that Clint Eastwood sometimes uses in scenes of threat: 'You know you're going to look awfully silly with that knife sticking up your ass.'

> 🍃 I identify about two-thirds of the birds I encounter from song and call. I learned how to do it mostly by listening to recordings. A birdsong app for your phone like Chirp! is useful and not expensive. You can also search 'RSPB birdsong identifier', which brings together the songs of common garden birds. It's all about making a start.

Sunday 4th

The wind improved the experience of my daily puddle by finding the sweet spot between my shoulder blades. That meant it was coming from the southwest. Wind is a problem for some birds but an opportunity for others: a female harrier crossed the marsh in front of me and then crossed back.

She had to work pretty hard as she headed into the wind, flap and glide, the gliding period shorter than her preferred rhythm. But then, as she turned her back on the wind, she made a completely different shape, becoming almost a completely different bird as she assumed an outline like a recurve bow.

In this form she took the wind under her tail and went skiing rapidly down a slope of air, crossing the marsh in about one-twentieth of the time she had needed for the journey out and losing no more than ten feet in the process.

I thought about flying as I sat with the wind strong in my back. The harrier's behaviour looked aimless, but there was nothing playful about it. Perhaps daily flying is a necessity. A musician plays scales and arpeggios every day; a marathon runner runs daily miles; a cricketer routinely works in the nets. Fitness: power: muscle-memory.

The acquisition of any skill brings its own reward: not just in successful hunting, but also in the pure gratification of performance. I can feel this myself when I paddle my kayak on the local river. But that harrier: the turn away from the wind, the power-slide that covers half a mile in a handful of seconds ... I would swap all my words for such a flight, for such a handful of seconds.

> 🌿 Finding the name of what you're looking at is only the start. The more you sit, the more you understand that it's not bird spotting, it's bird *watching*.

Monday 5th

I sat in sunshine; fifteen seconds later I was sitting beneath cloud; a minute more and I was sitting in rain that was confident, steady and in it for the long haul. It was unrelenting but I was unresenting: by now I had travelled too far to consider going back, and besides, I wanted to see what happened next.

What happened was herons. Two of them in the air. Few British birds give a better impression of hugeness than a heron in flight: at the same time laboured and powerful, big slow beats, each wing an arch. A harrier's wing looks like a bow when viewed from above; a heron's wing is like a bow from head on, one of those complex twin-arched bows.

The big grey wings have black leading edges with patches of white, like headlights. I wondered why. Perhaps when they enter a heronry – as these birds would be doing in a few months – the double-flash of white marks them out as herons, not, even for a second, to be mistaken for birds of prey, so they won't cause panic or prompt aggression from their neighbours.

The herons ceased flapping to descend: parachuting down. The heron in the rear touched down close to the distant river; the leader carried on a little longer; I almost fancied I could hear the thump of the landing. A little later a third heron flew over. Wings unhurried, headlights on full beam.

> The UK heron population of the early 2000s was the highest it had been for ninety years. The website for the British Trust for Ornithology (BTO) will answer all kinds of important questions.

Tuesday 6th

The seat was soaked as usual and its chill if not its moisture penetrated the fastness of my waterproof trousers, but there was a little blue in the sky and I had the briefest glimpse of a marsh harrier, a female –

... but how did I know? Am I genius or something? Let me come clean. I have been looking at nature for many years and I am ... well, not too bad when it comes to ID, at least of birds, though no top-flight observer would mistake me for good. I lack the precision of mind and gaze that top birders possess. And my knowledge of most other taxa is patchy, as these pages will show.

But I knew this was a harrier. The flight silhouette told me that: long body and tail behind the wings, wings two narrow rectangles held in a shallow vee, in this light the body appeared a uniform brown, the head the colour of clotted cream. Years of looking at harriers had created a harrier template in my brain: when I have a half-decent view of a harrier, I know what it is. The more you look the more you see: it really is as simple as that.

But when I saw a second marsh harrier I gazed with more doubt than certainty. This bird was subtly different to the female, but was clearly not a mature male, for they are very different. The head was less creamy, the underwing colour was not solid, more broken up.

So it was a young male. Perhaps two or three seasons old, not yet ready to start the challenging life of a breeding bird. I reached this conclusion after taking notes and consulting three field guides. Would I have troubled to do that, had my mind not been concentrated by sitting? Who's to say? A better birder would have known all this all along, but it's not a competition, is it?

A leaf fell from the ash tree above me: well, a few leaves all together. Get back on the branch, you little traitor.

> The field guide that clinched it was *Britain's Birds*, published by Wildguides. The advantage of a book over electronic resources is that you can thumb through, not only searching but learning.

Wednesday 7th

I sat in sunlight and the world was so shocked there was scarcely a bird in the sky. The view across the floodplain is lovely enough, and the contemplation of these empty acres was no hardship. But the worker, the writer in me itched for something to record, something to make copy. I sat on. Still nothing happened: and even that primordial instinct to write grew less urgent. Just sky and reeds and willows and the smallest touch of warmth from the sun.

A small frond of leaves detached itself from the ash and spiralled down, landing with immense precision on the seat alongside my own. Winter was coming.

And then, glory be, I saw a leaf rise up from the marsh and fly purposefully though the air before reattaching itself to the ash. It was clearly a miracle, time in reverse, autumn moving inexorably into summer. An illusion, of course, but a joyous one: a butterfly, almost certainly a red admiral, was playing the part of the leaf: a tough-guy butterfly bulling deep into autumn.

> 🍃 You get quiet days, and they are to be enjoyed for their own sake. It's best to set out with high hopes and low expectations. And you always find something.

Thursday 8th

A travel book should take you beyond your own experience: night-hunting with leopards, waiting on the convenience of elephants, crossing crocodile-filled rivers ... and I would have been doing those things if the world had been otherwise. Instead I put my right ankle on my left knee and leant back in my chair, the better to appreciate the rain.

Here was a seat less travelled by, a landscape seldom seen. A rainy day is not the same on the far side of a pane of glass, for it lacks the faint scent of rain, the soothing rhythm of the drops on your well-clad body, and above all, the knowledge that you are not observing but participating.

The hard edges had been washed away, trees and sky and grass and reeds softened and merged; the colours, viewed through the lens of rain, were gentler, less insistent. The landscape was a tone darker, from the sunlessness, and yet it was more luminous, filled with light from each refracting and reflecting raindrop.

Not for the first time, I wished I could paint. Did Monet have an umbrella when he painted his rainy day by the Seine? Vincent painted a rainy landscape at Auvers three days before he died and it's a picture to break your heart, and for the hope it shows rather than the despair.

We can all accept the beauty of rain, but mostly we let others explain it to us. Now I could see it, hear it, smell, feel it for myself, and so I took a lingering look at the alder carrs, the willow stump, the sallow bushes, the line of the river and the all-covering cloudscape of shining grey. It was like living on the inside of a pearl. Quite inexplicably, two great spotted woodpeckers flew over.

> 🍃 When you're out in nature you have something hardly anyone else in the country has at that particular moment ... and it's never too long before something reminds you of that privilege.

Friday 9th

A few pigeons, a few rooks, a few jackdaws. Not sunny, not rainy. No action of any kind. It wasn't a lot like a wildlife documentary. True, there was life and it was wild enough, and, for that matter, it was being documented – but it was all rather short of mega-events in the lives of megafauna.

Admittedly – though I bore them no ill-will – I'd be thrilled if a peregrine stooped from the empty sky to kill one of the three mallards flying low over the marsh, but that wasn't really the point. Wildlife camera-people wait months for the one killer shot: they don't show us the waiting, the quiet times that dominate even the most exciting habitats on earth. Perhaps that's why I was sitting here.

I scanned around, looking for the drama of a leaf changing colour. Then I picked out a patch of red on the opposite side of the river. I'd have sworn it wasn't there yesterday, but it must have been: and it was berries, not leaves: hawthorn berries standing out from the surrounding green: notice me, eat me, spread my seeds wherever you fly.

> 🍃 You never know what you will find ... and the process of looking is good in itself. It makes an explorer of you.

Sunday 11th

There was a fair amount of blue sky as I sat down but the clouds hastily covered it, like a bather grabbing a slipping towel. Soon I was back beneath the nacreous sky of Norfolk in the rain, a hint of warmth from the hidden sun behind me. I could hear the distant bell-beat of swans' wings.

And then like a conjuring trick there was a rainbow over the right-hand carr: Richard of York and Roy G. Biv in their joint splendour. In the space of a single instant the whole world was luminescent and lovely: and as if that wasn't enough, barrelling right through the centre of the great arch, two vast swans flew, so bright and so white it seemed they were lit from within.

Try to paint that. The choice of subject alone would condemn the artist as unforgivably sentimental. But I had to face the unacceptable truth: there really are moments of perfect beauty in the world: unsubtle, uncompromising.

It's been said that the last movement of Beethoven's Ninth Symphony, when the choir sing Schiller's *Ode to Joy*, is not based on the sentimental idea that life is actually like that. It's a glimpse of what life could be and should be and mostly isn't.

But here at the end of the rainbow, joy was made visible. When I came to write it for others to read, I would try to undercut the experience with wit, nuance, irony, allusion, understatement and humour. But right now I was fresh out.

> Mute swans are the world's second heaviest flying birds, after the kori bustards of Africa; a male mute swan can weigh up to 12 kilos (26 lb) with a wingspan of 2.4 metres – nearly eight feet.

Monday 12th

It's not actually raining but when you've said that you've said all. That's how my mother would have described the day. A robin sang to my right; and another struck up to my left: soft sad songs to match the day. The oaks on the far side of the river were beginning, just beginning to turn: a faint lessening of green.

Ah, the tranquillity of nature. Not so tranquil for the invertebrates in the scrub to my left: a wren was working the place methodically. Nature is no more tranquil than it is sad: all that stuff's just human fancy.

Then the sky filled up with ducks: perhaps 200 of them in small groups. They were mostly mallards, with some shovelers, from the head shapes and huge bills, and, from the speed of their wing-beats and their smaller size, a few teal.

I watched their unceasing restless manoeuvres, wondering what had disturbed them. Was it the buzzard I had seen earlier? Autumn is all about survival: getting through to spring and a chance to breed. To feed. Rest. Avoid danger. Save energy. Every time a bird gets disturbed, it flies about burning hoarded energy. Too much disturbance and the bird outflies its fuel supply and dies. That's the problem with shooters: it's not the few they kill; it's the many they disturb.

There is always a fair amount of fretting to do when you engage with the wild world. It's the price you pay for loving the stuff.

> 🍃 Identifying birds on the wing is hard: they don't keep still like they do in the book. Studying the pictures helps: idle thumbing through can pay later dividends. If you place a field guide by a favourite (indoor) sitting place, you can do a few minutes of mugging up any time.

Tuesday 13th

Once again I was travelling through the country of the rains: a place where I was no longer a stranger. Two mallards flew past: I checked them through the bins. As usual, the female was in the lead.

People come to Norfolk for cycling holidays, relishing the challenge of our mountains. In summer I often see male–female couples on tandems. I have never seen the woman at the front. Cindy said the arrangement allows the woman to take a break from heavy pedalling unrebuked.

And then I heard the sound of beauty: not a beautiful sound in itself, but a sound whose meaning is beauty. It was the *whut-whut-whut* valley-filling sound of swan's wings: the bell-beat of their wings, for I borrowed that phrase from Yeats: far-carrying, rhythmic, hollow. A pair of them was flying the line of the river half a mile away. Couldn't see which was in front, but it was either the male or the female.

> 🍃 In the colder months it's always worth giving swans a second look: whooper and Bewick's swans fly in from the Arctic to enjoy our balmy winters.

Wednesday 14th

I sat down in a big wind. Hurrah! A new kind of weather to write about. Three jackdaws rode the gusts with playful insouciance. Then the wind brought rain, a sudden violent squall, and I was back in familiar territory: bum cold from the remains of the puddle, hands very cold on the binoculars, face stung by the fierce water.

Bad weather isn't bad. Not if you have a home to go to, anyway. And certainly not when you're committed to it and have the right gear. After all, this was good fresh water; the stuff of life. Though perhaps the bumblebee saw it another way, zagging round my ankles as if trying to dodge each individual drop, knowing a serious soaking was death.

How long is a proper sit? Can I go in yet? After a while you find yourself moving through the restlessness. Soon enough you're just sitting.

Movement on the marsh, low down, just above the vegetation, big. It's always movement that catches your eye: it dropped long yellow undercarriage and extended a snake's neck: heron, then. The weather was not going to put him off his fishing: he looked as if he was wearing a huge dirty mackintosh as he commuted from one dyke to another. If he could cope with this, so could I. But put gloves in your pocket next time, right?

> Fingerless gloves are best for wildlifing in all but extreme weather: easier to use your bins and scribble in your notebook.

Thursday 15th

Even as I was tipping a relatively small puddle from my seat, I heard a green woodpecker. I've been living with nature too long for it to sound like mocking laughter: it sounded like a green woodpecker. Is that my loss or my gain?

I sat as the chill but not the water penetrated my waterproof trousers: I was an old hand at this now, or perhaps an old bum. I no longer expected to be rewarded by an especially good bit of nature.

I got one anyway. The green woodpecker flew into sight, up-down, up-down, and slapped himself onto the old willow-stump. Once there he made a rather perfunctory search for invertebrates buried in its fibres. I could see that it was a male from his red moustache.

Quite a bird: livid green against the pale trunk, the red marks on the bird's head brighter than the distant haws. It was almost as paradisial as the jay. I noted the dagger-bill, the fierce eye; he noted nothing of me in return. My stillness had made me invisible.

Then he was off, revealing a great sunburst of yellow on the retreating bum. What a day for bums. It rained a little after that but my waterproof trousers and I didn't mind.

> Watch out for that up-down switchbacking flight: all our three species of woodpecker fly that way. The call – the yaffle – of green woodpecker is very distinctive once you've got the hang of it.

Friday 16th

Almost frighteningly loud, the voice of a green woodpecker in the lower branches of the ash. Fewer leaves than yesterday but I still couldn't see it. Perhaps it really was mocking laughter this time.

There was a minute or so of sun, quite a treat. The clouds were high and thin and there was even a little blue sky. Anyone could do this, I thought. Admittedly that was (and is) the point.

This shift in the weather had made the whole place subtly busy: making up for the time lost to bad weather. A group of goldfinches called in their pretty tinkling way, and eight of them flew past like a collection of yoyos. (Goldfinches bounce; woodpeckers undulate.) A different or the same green woodpecker called from the meadow to my left; a marsh harrier caught the sun and for a moment the pale head was ablaze.

Then a brief exchange of views from two little owls, cheerful yelps that seemed full of confidence. Sustained wet weather is not good for owls; they can't hunt with soaking feathers. They were relishing the let-up, however temporary, and so was I. There was even another little burst of sun.

> The marsh harrier was either a female or a juvenile: not easy to tell apart without a good view, as we have seen. In a good few species the young of both sexes look like females: that way young males aren't seen as rivals by mature males.

Sunday 18th

No bravery needed today, not once I had made the decision to sit in what was left of the daily puddle. It was almost balmy. The distant call of curlews came within a couple of minutes.

This is a project that must celebrate the ordinary and the daily: black-headed gulls, wood pigeons, jackdaws and rooks. You can't spend your life writing about the elite; Homer wrote about swineherds as well as gods.

So I listened to the sounds coming from the rookery behind me, wishing I knew the meaning of each one of their many different, surprisingly various calls. Then I heard a rook-call that was out of the ordinary, looked up and saw one bird chasing another, calling hard. The calling bird caught up with the other and performed a 360-degree roll all round it. It was quite a feat of aerobatics and seemed to be a kind of non-tactile aerial embrace.

Rooks need to be part of a flock, but at the same time half of a pair: two among many. This barrel-roll might be the first coming-together of two young birds; it might be a brief moment of affection between the oldest pair in the colony.

Some people think rooks are dull birds, but I suspect that's as much an oxymoron as bad weather. Six curlews flew overhead with a light scattering of fluting calls.

> This sitting business – it's the difference between seeing a great picture in a gallery and hanging it on your wall: you're always finding something new in the deeply familiar.

Monday 19th

There was no puddle on my seat, instead a fallen frond of ash. The seasons were marching on.

Movement. Harriers' eyes, deer's eyes, butterflies' eyes: all are drawn to movement, as we have already seen. Two important matters come from this. First, if you can stay still, you have a cloak of invisibility, drawing no eye. Second, your own eyes are always ready to respond to movement.

I followed the movement with naked eye, a bird recognizable as a marsh harrier before I lifted the bins. As I focused it stopped and perched on a sallow bush across the marsh. Perhaps it was a favourite perch, one I had never noticed before. After a few moments the bird took off and flew straight towards me, so that I could appreciate fully the shallow vee of the wings: wings held at, I estimated, 150 degrees. That wing profile is a dihedral, and it gives the bird stability in the rolling axis, along with a tendency to return to level flight in disturbed air. Most of the big airliners use the same idea, though birds thought of it first.

A single goldfinch flew over. Another movement in the sky: and my eyes followed it almost without volition: a newly abandoned frond was spiralling down from the ash.

> Ash, sallow: it's good to know trees by name, and not hard. Good, straightforward information can be found on the Woodland Trust website; they do a phone app as well.

Tuesday 20th

Looking and listening are different activities from seeing and hearing. They are active states; questing and dynamic. They're not contemplative, but they're quite good if you want to find wildlife. I sat in the warm sun on my puddleless seat. The marsh harrier perch was vacant. I saw a buzzard on the far side of the river, and heard a Cetti about 300 yards off.

Perhaps I should perform these sits without binoculars, surrendering myself solely to the pull of the earth. After all, Gilbert White had to rely on his naked and acute eyes. But I sat with binoculars at the ready, in case the black-headed gulls along the river should turn out to be something else.

And that white bird was no gull. It was a little egret: brighter, more rounded and yet more delicate in flight. The light was so good I picked out the yellow feet on the end of dark legs: the birds will waggle these at fish as a lure.

> Gilbert White first published *The Natural History of Selborne* in 1788-9, and it has never been out of print, which is something to aim at. Reading the best books enriches your experiences in the wild and gives them depth.

Wednesday 21st

I am the thorn of an unkind friend. A line from the Incredible String Band. Light rain was falling, a puddle on the chair. Three parties of ducks were crossing and recrossing the marsh: two of them, flying in tight formation, were fat-beaked shovelers. A duck's flight is always urgent, hectic, hurried: they know no other rhythm.

A little owl called. So did a car alarm, rather closer – an unusual record for round here. The car, clearly distressed, made its alarm call at unpredictable intervals, starting up again just when you thought it had settled down.

I tried to compose my mind. Four greylag geese flew over, heavy and purposeful, a slower rhythm than the ducks. My mind kept returning to a friend's unkind email. I wished to get both him and it out of my head, but neither was in a hurry to go. The other evening I had cleared about fifty burs from the pony's mane: the unkindness was like the burs, inextricably entangled with the stuff of my head.

I tried again to ease my mind and allow nature to do its healing work. The car gave out another alarm call and I gave up. It'll be better tomorrow. Or if not, at least still here.

> The more you give nature a chance to make things easier, the better a job it is likely to do. And even if there's a car alarm going, you can always go back the next day.

Thursday 22nd

An ash tree lets go of its leaves in clumps, the stalks performing a brief spiral on the way down. It looked as if the wind was too impatient to pick out single leaves, instead ripping them off in handfuls. A look at the leaves on the empty seat beside me told the true story.

I could see quite clearly the point at which the tree had not lost its leaves but forced them off. It's a process called abscission: the tree selects a point at which it denies nutrients to its light-gathering devices – leaves – because it's not economically efficient to keep them in the light-poor days of winter. Autumn is not the fall but the push. Soon the branches above me would be bare. The clocks would go back on Saturday.

But at least there was a little sun today. A heron made a long glide from the heronry to the river, a beautifully judged semicircle. Just to show this was no fluke, a second heron did exactly the same.

My feelings about my unkind friend were already changing. Soon distress would turn to anger, anger to irritation, irritation to a weary rolling of the eyes. In the fullness of time we would drink and talk together as always. I leant back and a great thought crossed my mind. Fuck 'em.

Nature heals, and there are times when the thought 'fuck 'em' is as soothing as prayer, for after all, it represents a profound philosophical acceptance of the way things are. A buzzard rode the wind with immense aplomb and gave out a great butine yowl. That's right, buzzard. Fuck 'em.

> Why and how are always good questions: seeking the answers takes you deeper into nature. Trees stop making chlorophyll in autumn because they stop photosynthesizing; that's why they stop being green.

Friday 23rd

We've all seen time-lapse footage of clouds, heaving and shapeshifting across the sky. Today it was like that in real time: low clouds in fifty shades of grey, silver and black: a different sky every minute. Then there was the tiniest bit of blue and a hint of warmth on my back. In such strange light a strange illusion occurs.

A black-headed gull flew past, as ordinary a bird as this piece of sky can offer. But this was not an ordinary moment. With the dark clouds behind it and the hint of emerging sun before, it seemed that the source of all the light on the spreading floodplain was the bird. It glowed like a star; it descended like an angel bearing good tidings. A black-headed gull is a canny bird that knows at least fifty ways of finding food: but it was as if I was looking instead at the albatross that guided the ancient mariner through the icebergs.

It flew with shimmering wingbeats across the marsh – and then the light changed and it was a gull again and I was a bad birdwatcher sitting on a wet chair. A little later, forty redwings flew over, bringing winter with them.

> Science is not the only way of enjoying nature. Art and science both inform our understanding and enrich our pleasure in nature.

Sunday 25th

How many man-made items could I see? The tip of the mast of a boat moored on the river, dead in front, the boat itself hidden, and to my right, the top of a pylon. But I could hear a few: the main road is about a mile away behind me, the apex of a long, fast bend. That makes it a thrilling place for Sunday afternoon motorcyclists: like coloratura sopranos, they constantly strive for the highest note in their range.

But I was in good heart, gave the repeating sounds little more than an eyeroll, and thought about the famous high C in Allegri's *Miserere*: when the music itself soars like a bird. The wind was behind me, but the day was almost warm, and I could hear a pied wagtail in the meadow to my left. Probably two of them; there usually are.

There's something about a buzzard that makes them seem huger than they are: something to do with the depth of their wings, and the resulting floppy, almost clumsy nature of their powered flight. They lack the well-rehearsed neatness of harriers: they seem to be making it up as they go along.

A black-headed gull flew over. It looked exactly like a black-headed gull.

> Black-headed gulls only have black heads in spring and summer. You can tell them apart from other routinely seen gulls because they are slighter and neater, with a bright white leading edge to their pale grey wings.

Monday 26th

Nice timing. Five minutes after I had taken my seat a violent shower began. It felt quite different to the rain of a few weeks back: all the water was now coming directly from the sky. The ash, all but leafless, was no longer able to hold the rain and release it in big drops. I looked up: the sky was almost ostentatiously visible.

The rain, I noted, was neither pleasant nor unpleasant. It just was. This seemed to me an advance, even if I could have done with the gloves I kept promising myself. Just part of the landscape, that's me, I told myself vaingloriously.

The wet sky was busy with jackdaws, crossing the marsh in a series of straggly pairs. Jack, one would call to other; jack, came the reassuring response: like an answering smile from your beloved.

The rain began to slacken off, and a small bird flew from one of the sallow bushes. Great tit? And then another. And another. In all, two dozen birds flew out in a loose flock, mostly great tits and blue tits. Unlike me, they had been sheltering, the wimps. Now they were off to forage together, once more putting their faith in safety in numbers.

> Ethologists – students of animal behaviour – call the stuff those jackdaws were up to 'reinforcing the pair-bond'. Human pairs do it all the time: and it's no great mastery to observe it in other species.

Tuesday 27th

The day was balanced precariously on the edge of rain. I sat down and looked at pigeons. Then my phone rang. A second later there was a shattering of twigs in the bushes ten yards away, followed by the sounds of a retreat through the crackling vegetation. In Africa I'd have jumped a mile; here I just smiled. Chinese water deer. I wondered how often and for how long the same animal had sat tight while I did the same. Had it sat with me through the showers?

Perhaps I should have chosen a less strident ringtone. I felt rather embarrassed. A favourite bit of birdsong, perhaps? If I kept this project going for a few more months I'd be sitting here overwhelmed by the song of birds. Real birds, not ringtones: wild thought.

Another voice lifted up in song: and it was Eddie enjoying a musical session in his room 100 yards off. He sings as Algernon plays the piano in *The Importance of Being Earnest*: 'I don't play accurately – anyone can play accurately – but I play with wonderful expression.'

There are many enjoyable books about wild quests: to see 200 bird species in a year, or every species of British butterfly, or to visit every English county. My quest was to spend a year not looking for anything: the questless quest. No, it wasn't. My quest was to find a nice pair of gloves.

> The quest books mentioned above, all worth a read: *Why Do Birds Suddenly Disappear?* by Lev Parikian; *The Butterfly Isles* by Patrick Barkham; *Engel's England* by Matthew Engel.

Wednesday 28th

Blue sky with scattered cumulus clouds; above them, a line of wispy cirrus, and beneath them, an observer wholly out of sorts, grousy and grumbly. It had been that kind of day. Cirrus are sometimes called mares' tails, and it's a dirty joke: a mare will pull her tail to one side when presenting herself to a stallion.

As I scanned the green expanse before me, I picked out a tiny patch of white a mile away and knew it for a swan before I had raised my bins. The bird was dozing in plain sight. Unlike yesterday's deer, it was utterly confident in its size and strength.

I sat on. This is not exactly meditation, but it's not exactly birdwatching either. Sometimes a string of negatives is the only way to get close to a concept: not watching, not dozing, not eager, not resigned. When you try for exact definition the entire concept disappears, like taking too firm a grip on a bar of soap.

Blake would find the search for definition footling. He recommended that you kiss a joy as it flies. Not that this was exactly a joy either, but – well, this trying day was rather less trying by the time I was ready to move on. Not joyful, not grumpy.

A kestrel flew past, unswerving. He who blows a kiss at a kes as it flies dwells in eternity's sunrise.

> You see more nature when you go out and look for it. Every day of this quest made that clearer. Changing habits changes everything.

Thursday 29th

The quest was over. I sat warm-fisted in the rain, black gloves on my hand. I had found them in my town-going coat, exhumed for winter. The index fingers each bore a touch-sensitive pad, so that I could use my phone without removing my gloves: didn't they know phones were against the rules?

Who'd choose to fly in this weather? Only a few black-headed gulls, getting on with stuff over the river. And of course a couple of wood pigeons. Perhaps this year will bring me a day when I am enthralled – or even mildly intrigued – by wood pigeons. If so, it has yet to arrive.

But I'm not looking for rare birds, am I? Certainly this would be a capricious way to do so: you find rare birds by running about, actively chasing stuff. Perhaps I should be looking for rare bits of myself, but I wasn't doing that either. I was just sitting in the rain, hearing a blackbird call briefly, noting two swans flying in tight formation along the river.

The horses in the next field started; there was a sudden scrabbling in the bushes, and the deer was moving again, but not because of me. I was silent, I was still – I was the rain, I might say, if I was a Zen monk. Certainly I was the rain's target.

> If you want to look for rare birds, you can find plenty of help. At the big nature reserves there's usually a whiteboard with a list of what's around. There are also plenty of websites with details of local and national rarities. Rare Bird Alert operates a subscription service.

Friday 30th

A dry chair, a hint of sun, almost warm: why, anyone could sit out on a day like this. I was doing nothing special. I had a stupid moment of disappointment. The sky was busy with gulls and corvids, enjoying the wind that was chasing the clouds along. The air above the marsh was busy with insects: midges, dayflying moths, even a bumblebee.

A small group of fieldfares flew by on the far side of the river; their black tails made them obvious. They'd have their eyes on those red berries; there were still plenty on the tree. These are thrushes that join us for the winter from Scandinavia, as if to remind us to look for the best in dark times.

Then my father rang up; no deer fled at the phone's summons. The general prohibition on phone use doesn't extend to him: he had managed to walk 50 yards from his house and back and was in good form. We talked of a 60-mile walk we had done a few years back. Owing to a miscalculation we had to do 25 miles on the last day, though perhaps today's marathon had been even more taxing. We discussed in detail the pint we had drunk at the finish.

> Britain's betwixt-and-between climate means that many birds leave us when the summer ends, but a good few come here from further north to savour our gentler winters.

NOVEMBER

Sunday 1st

A dry seat, a hefty wind, no sun but a brighter sky to the north. And then a shaft of piercing beauty, one that made me speak out aloud in my astonishment. I think all I said was 'marsh harrier', but I spoke with wonderful expression.

Jackdaws love a big wind, and there were fifty of them

in a sprawl across the sky, rising, falling and spinning with untidy but effective movements of their wings. Right in the middle of them I picked out the harrier, and there was something unspeakably wonderful about it: the contrast between the busy, ragged silhouettes of the jackdaws and the still-winged, easy glide of the harrier: corvid and raptor, generalist and specialist, ordinary and special, prose and poetry.

It was like finding a diamond in a jar of pebbles: but this intends no disrespect to either pebbles or jackdaws or, for that matter, prose. It wasn't that one was better than the other: it was that they were thrillingly different: like the prose and haiku mixed together in Bashō's *The Narrow Road to the Deep North*. And that's nature for you: everything thrillingly different from everything else. These days it's called biodiversity.

The jackdaws surrounded the harrier but made no attempt to harass it, and then the jacking ball of corvids was going left while the lone harrier maintained a glide in the opposite direction. Wings held high.

> As you watch birds (or anything else) you become more and more capable of processing limited information in a meaningful way. It's that template business again.

Monday 2nd

The ash tree was now almost bare. It cast a shadow but made no inviting shade, but there was at least sun, at least for a while. Two buzzards were calling to my left.

A herring gull cruised over the river, not an adult, a second winter bird, as far as I could make out. But he was old enough to possess the gull's traditional air of nonchalance: of achievement without effort, like a lordly batsman scoring a languid century.

The sky was in teasing mood. First it blacked out the sun, leaving me a little downhearted. But followed this with a sweep of gold: a newly gilded landscape rushing towards me, transformed inch by rapid inch by the advancing sun. It was like a lover running to embrace me ... in *Ulysses* 'quick warm sunlight came running from Berkeley Road, swiftly, in slim sandals, along the brightening footpath. Runs, she runs to meet me, a girl with gold hair on the wind.'

> The plumage phases of gulls provide endless fun on a quiet day. It can be frustrating, but I tell myself that nature is *supposed* to be complicated. Isn't that why I'm here?

Tuesday 3rd

For the first time the cold really bit: the hard, dry cold of winter that's quite unlike the cold of a rainy day. It seemed to have taken the world by surprise: nothing about but a few corvids, a handful of pigeons and a couple of black-headed gulls.

Only the oaks were holding onto their leaves, and even these were thinning. Most of the trees were now stripped bare, having pushed their unwanted leaves out into the scurrying wind. The colours before me were changing by the day.

Yesterday I had thought of *Ulysses* and Mr Bloom; today I thought of Modesty Blaise: another eternal favourite. As I wondered about definition by negatives I remembered Sir Gerald Tarrant trying to define Modesty's bearing when they were both prisoners, escape impossible, death certain. 'Negatives were easier. Not defiant, not optimistic, not grim. Just totally absorbed by the problem.' I sat on, not excited, not bored, not happy, not sad. And certainly not warm.

> An ultra-lightweight fleece makes a very good additional layer on such days. Not heavy, not restricting, not inconvenient, not even very expensive.

Wednesday 4th

A clear night sky had brought the first frost of the year and a crisp white morning. A day of unbroken sunshine followed, and its warmth surprised me as I took my seat.

A party of gnats danced in the air before me: nothing fanciful; this was a dance all right. The males take advantage of these days of unseasonal warmth and form gatherings in the air, rising and falling, confident that no swallow would be there to devour them. It's all done in the hope that a female gnat will join the party. The dance of the seasons seemed suspended in this unexpected warmth. I felt as if I had been excused time, as the malingering soldier is excused boots.

A flock of sixty ducks came into sight from behind the heronry, presumably from the Flood. They were trying to make the disciplined peloton formation that ducks love, but were too agitated to do it properly. From the pale belly and the white wing patches on some of them – the males – it was clear they were wigeons. They crossed and recrossed three or four times, round and round, no leader, no purpose other than a need to work off their distress.

There had been gunfire at dawn, but nothing more for eight hours. What had upset them? Eventually they disappeared behind the alder carr on the right.

> The best way to count wildlife is in blocks. I counted the wigeon in tens; you alter your order of magnitude to suit the numbers before you.

Thursday 5th

Another bright clear day: high above hung the mackerel clouds called cirrocumulus. Hard to think of them as ice crystals, sitting below them in the sun. A great spotted woodpecker made a long flight, at least by woodpecker standards, from the heronry to the carr on the right: their undulating flight always looks such hard work.

A buzzard called – two buzzards, bursting from the same right-hand carr, big wings flapping hard. Buzzards also seem to find flight hard work before they have gained some real height and can glide. They look so untidy for birds of prey, until they can stretch out their wings and relax.

Were my thoughts on yesterday's distressed wigeons a little presumptuous? It's a dangerous area: I can't exactly prove that wigeons have feelings. But we have surely moved beyond Descartes and his assertion that non-humans are merely clocks that respond to a stimulus. I tried to remember a line I had read on this debate: to say that a tortured puppy doesn't feel pain is a remarkable achievement 'even for a philosopher'.

A robin sang a few laconic phrases from the thicket where the deer had been hiding.

> The ethical philosopher Peter Singer's work of 1975 *Animal Liberation* is very good on this and related issues, also Mary Midgley's *Animals and Why They Matter*.

Monday 9th

It was good to be back. Various professional and family concerns had kept me away from the seat beneath the ash, but an enforced break makes the routine all the sweeter when you return: low, dark clouds and a marsh harrier still darker over the woods on the far side of the river. The bird was a mile away but utterly distinctive: two simple brushstrokes from the master calligrapher are all that's needed to spell out the bird's presence on the great scroll of the Norfolk sky.

There's a privilege in familiarity, in taking certain things for granted. The point was not that I was clever enough to recognize a bird from quite a long way off, but that I had lived with these birds for so long that they had quite literally become a part of me: their image was lodged in my brain, which now responded in recognition to the smallest clues given by a harrier. The other day, paddling my kayak, I had recognized a harrier from its reflection on the water.

Four swans flew along the river in close formation, a

slightly grubby sub-adult at the rear playing catch-up. Then a harrier flew almost directly overhead at about 30 feet ... and I was aware that there were two of them, both straight ahead: one now at about 20 feet and the other 50. One had a very creamy head, the other less so, and overall much darker. The first was an adult female, the second a juvenile. My brain still doesn't know enough about harriers: I wished I was better at making this distinction at distance. The two birds manoeuvred around each other for a while, each very much aware of the other, an exercise that no doubt maintained the parameters of their relationship. And then I was alone with the sky again, me and a few black-headed gulls.

> Learning to respond to wildlife is essentially a process of changing your brain. That comes about through habit and will.

Tuesday 10th

When the story you're telling is natural history, you have to accept that many narratives will lack a satisfactory end, and a good few will also lack a beginning. It was a clear, cool day, the sun giving the palest tint of orange to a few cumulus clouds. My sisters wore colours like that in their childhood dance classes; a dance-form called Natural Movement.

And then it happened. Except that maybe it didn't. A natural movement in the air above me. Two birds – I think there were two – in my peripheral vision. I had one of them in the bins for maybe a quarter of a second: pigeon? Falcon? Did I see the hint of a moustache?

They had gone into the big ash just above me. I forsook my chair – recklessly breaking rules – and circled: with trees, a six-inch shift can give you a completely different view. Surely I saw a tail with a dark terminal band. Or did I?

At last a movement. I caught it in the bins – and it was a goldfinch. Delightful, sure, but not the bird I'd been looking for, and certainly not the bird I had seen going in.

So what was it? I'd been having wild thoughts of merlin, a tiny falcon seldom found in Norfolk. Birders grade doubtful sightings as prob and poss. This was a notch lower than poss: perhaps fant for fantasy.

But this wasn't a matter of pure frustration. There's beauty in incompleteness, the surviving fragment of a lost poem. The blue sky seen through the ash reminded me of one of Vincent's almond-tree paintings. Two shelducks flew downriver.

> Birds can fly. Well, most of them. That makes them great wanderers. Almost any species can turn up almost anywhere at almost any time. There are two species of albatross on the British list. No ostriches, though.

Wednesday 11th

The sky was murky and so was I. Trying day. The sit had assumed the status of a treat, and I even had a little chocolate to mark the occasion. So I sat and munched and things got a fair bit better. Then, infuriatingly, I dropped the last piece on the ground. I would have eaten it – just good clean dirt – but

I couldn't find it. New discovery: fallen ash leaves can take on the colour of milk chocolate.

A flight of thirty thrushes flew over: a wintry sight, almost certainly redwings. There was a busy crowd of 100 gulls over the river, circling, taking advantage of what seemed to be a rather feeble thermal. It didn't allow them to gain much height, but they were able to cruise around using very little energy, apparently just enjoying the company.

I picked out a marsh harrier in the middle of them: as black against the sky as the gulls were washing-powder white: good clean dirt, perhaps. I focused on the harrier and caught the smallest adjustment of the aerofoil surface and the angle of the wings, and the bird has changed direction, moving out to the far rim of the circle of birds.

I rose to go: a heron rose with me. It had been fishing just 20 yards away, unseen. I apologized out loud. Then I had one last look for the damn chocolate. No luck. Consigned to the economy of the soil.

> A thermal is a patch of air warmer, for various reasons, than the air surrounding it. So it rises. That means free lift for birds that can exploit it: gulls, corvids and birds of prey all like a good thermal, which makes thermals good things to watch. The birds will find them for you.

Thursday 12th

It was cool and sunny and I was full of that contented body-weariness that comes from good exercise, in this case a kayak trip along the river. From my seat I looked at its banks with a friendly, almost proprietary air. I remembered my resolution to check the harrier perch I had found the other day – and bloody hell, there was a harrier. This was quite absurdly gratifying.

It took off at once, as if aware of my scrutiny, though it was 400 yards away. I watched the bird as it made a low semicircle before dropping out of sight. So this was a hunting perch, not a resting perch: it had marked down a prey item, approached it from the blind side and, since the bird didn't get up again straightway, it had probably been successful: death in the grass.

Marsh harriers went extinct in this country in the nineteenth century, shot out by gamekeepers. The numbers climbed back after the First World War, but they were hammered by the pesticides of the 1960s and by 1971 were down to a single pair. Their revival gives us all a straw of hope to cling to.

They used to spend the winters in Africa, but now, in a changing climate, many of them stay. I sat there watching three of them, one gaining height very slowly in a thermal, one very low, just over the reeds, and a third above the woods on the right. All juveniles or adult females.

From a distance came the sound of a curlew, and I remembered the words of the Reverend Eli Jenkins in *Under Milk Wood*: 'Praise the Lord! We are a musical nation.'

> Pesticides like DDT got into the food chain and were inadvertently consumed in the greatest concentration by birds of prey, who eat the insect-eaters. The poisons made the shells of their eggs too thin to be viable. Many species of birds of prey have since recovered; making a sharp contrast with the general pattern of universal decline. We now use still more devastating and dangerous insecticides, but these don't affect birds of prey directly. They just kill insects indiscriminately, meaning that we have fewer pollinators, fewer insect-eaters and less robust ecosystems.

Friday 13th

The price of being a prey species is constant vigilance, but there, caught in soft autumn sun, the Chinese water deer was as relaxed as a deer could be. For this fellow, with his protruding canine teeth well on show, vigilance was more a matter of convention than practicality. No wolves round here. No tigers either, as his East Asian ancestors would have known. He was standing at the foot of a large sallow bush 200 yards off, browsing thoughtfully, giving the twig in his mouth a good old yank.

I caught a movement just in front of him, and with it a rich yet subtle flash of orange. It was a bird, jumping onto the plants that emerged from the low tangles of vegetation and then dropping down again: hunting from a perch like a tiny version of the previous day's harrier. Then a moment of stillness, the bird perched good and prominent, and I could easily make out my old cock linnet, not exactly a rarity but a nice view of a nice bird. There's a rich and subtle pleasure in long-range identification: and it's not precisely – negative definitions again – that of puzzle-solving or list-lengthening. It's about making a connection with the world: one way, yes of course, but still packed with meaning.

Linnet and deer moved into cover and I looked in some content at the late afternoon marsh spread out before me. And then a bullfinch began to sing, soft, secretive and sibilant. All song is a public act, including singing in the bath, but the song of a bullfinch always seems to be rather private, as if the bird really didn't want to be overheard. Perhaps this was a moment of private practice, inspired by the day's sunshine: an optimistic anticipation of spring. I, the eavesdropper, was happy to share it.

> There are times when a bird less positively flaunts itself, and I always try and take advantage of them: for the pleasure of the moment, yes, of course, but also as a long-term investment: learning a bird, getting its colours, its movement and its nature into your brain.

Monday 16th

Three swans were powering towards me from a mile off. I thought they would fly right over me but as they hit the river they turned sharp right, my left: two adults and one sub-adult: two extravagantly white and one rather grubby.

The young bird was lagging behind the others. The point of flying in formation is that the birds behind the leader meet less wind resistance, on the peloton principle of the Tour de France. If you keep the formation tight you get something of a free ride: lag behind and you must fight the air as hard as the leader, and it gets harder and harder to catch up.

That's not a problem for these mute swans who won't ever move far from this watery landscape, but the migratory swans – whooper swans and Bewick's swans, I've been looking out for them on the marsh every winter for years without luck – must keep up with the flock or die. The Tour de France has a vehicle at the back of the race that sweeps up the losers and the lost: the *voiture balai* or broom wagon. The wild swans that sweep in from the Arctic have no such luxury.

A few minutes later the three swans came back, retracing their flaps, working their way downstream, the youngest bird still three lengths behind.

> Both these species of less familiar swans have yellow patches on the beak. If only a Bit of the Beak is yellow it's a Bewick's swan; if nearly the WHole of the beak is yellow it is a WHooper swan.

Tuesday 17th

I had a cushion. It was designed for camping. It was so light I thought the package it came in was empty. When I opened it, the cushion looked useless. But I tried it anyway, on a grey, windy day; rather grudgingly, I noticed that my bum was a little warmer than it was on the naked seat.

The wind was a little south of southwest, gusting at 30 mph

so enough to notice. I saw a marsh harrier riding it, crossing the marsh with scarcely a flap, but with constant adjustments of its aerofoil surface. The wing shape changed again and again, the tail half-fanned and then narrowed; once or twice it made a half-flap to maintain its position relative to the wind.

The harrier was flying slightly north of east, or at 90 degrees to the wind. I remembered landing on Fair Isle in a crosswind of 50 knots; you don't always want to be so vividly aware of your pilot's skills. Here I could appreciate the same skills with less anxiety. I stretched out my arms and made a small series of flexions and tensions, wondering what it feels like to manage the wind so well, my body and mind filled with flight envy. Well, what else are words for?

Later I typed up my notes. When I checked them over I noted that instead of typing 'I saw a marsh harrier' I had written 'I was a marsh harrier'. After a moment's hesitation I made the necessary correction.

> The cushion in question bears the impenetrable message WWAGO. It was very cheap and fits easily into a pocket, extending the possibilities of a sit.

Wednesday 18th

A goldfinch flew over my head, calling enthusiastically. Same weather as yesterday: dry, sunny, gusty. The wind of recent days had been hammering the oaks across the river, but they weren't bare yet.

A buzzard flew round from behind the right-hand carr,

perched there for a short while and took to the air again, making a quite different silhouette to yesterday's harrier. Buzzards use the dihedral at times and for the same reason as harriers – stability – but it's not their habitual mode of travel. Mostly they hold their wings flat.

When their big broad wings are spread out and their tails are fanned to the maximum, they have a huge wing surface, and it helps them to soar; that is to say, to gain height with still wings (flapping is energy-expensive). They look for food from a height; harriers work much lower. Different skills, different wing shapes, different ecological niches, different species.

A harrier appeared below the buzzard to emphasize that existential point. The two birds worked around each other for a minute or so before going their separate ways; one high, one low. It was like opposite pages in a field guide – or perhaps an eloquent lecture on diversity. Why are there so many species of bird? Because there are so many different ways of making a living.

> Beginners do best with a field guide of modest ambitions, the fewer species and the smaller area it covers, the better. Example: *Collins BTO Guide to British Birds*.

Thursday 19th

The wet seat of my chair gave me a thrilling opportunity to give my cushion a more searching test. It was damp and windy with a bone-rotting chill: conditions that made my mother suspect personal enmity from the weather.

But heighho, I heard a buzzard call and caught a distant glimpse of marsh harrier, and it's a well-known fact that a harrier a day keeps the psychiatrist at bay. Nature is supposed to be good for your sanity, and even if I am not the finest example of that principle in action, I'd be much madder without it.

The benefits of nature for human health, mental and physical, are much talked about these days. I suspect that these benefits come most fulsomely to people who aren't looking for them. Nature is more generous when you go looking for harriers rather than for personal wellbeing. It's all in *Alice*: the harder you try to find the right path, the more elusive it becomes. Set off in the exact opposite direction and you soon reach the place you're looking for.

Anyway, it was a nice harrier. Cushion wasn't bad either.

> Two useful things to read: *Alice Through the Looking Glass* by Lewis Carroll, and Roger Ulrich's 1984 paper in *Science* magazine, 'View Through a Window May Influence Recovery From Surgery'.

Sunday 22nd

I've got used to it now. I was even anticipating it. I sat there on a cold, gusty day, the morning sun all gone, feeling restless, the weight of many tasks, professional and domestic, heavy in my mind. What to write? What to cook? How much longer was I going to sit here?

And then, slowly, like the sun coming out on a misty day, I would find myself sitting there in perfect comfort and content, shorn of all anxieties about what happens next. How does this happen? The process is elusive: suddenly you notice that you've stopped fretting. You've gone through.

The previous day, spent like most Saturdays with Eddie, had been full of wild incident: a hunting barn owl, a peregrine perched on the bench by the pond, two herons having a punch-up and, at dusk, a flight of 2,000 pink-footed geese. But now I sat unflustered with nothing to watch but a straggle of jackdaws flying together in the wind, about forty of them, no peloton dynamics for them. They were followed by another group of twenty, and then another a fair bit higher, all of them heading north, no doubt to the corvid roost just over the horizon.

A buzzard called. And then a strange illusion: I noted three ducks flying across the marsh – and then realized it was actually a single herring gull much closer than I thought: two wings held in shallow arches and the stout body between them adding up to three shapes, and my brain processed the information all wrong.

> Here's a rough rule for sitting out in wild places: the first five minutes feel like an hour; the next hour feels like five minutes. It's all about going through.

Monday 23rd

I work in a hut at the bottom of the garden – books and journalism, that sort of thing. It overlooks the marsh, so there's no excuse for ever doing a stroke. I had taken to keeping my cold-weather and wet-weather gear there. As I was putting it all on I saw through the window a male deer standing statue-still: in the alert posture, every muscle tensed. His head was turned away from me, so clearly I wasn't the problem. I looked along his eyeline and found another deer 200 yards off. After an extended pause the first deer went into a high-elevation trot: throwing a big chest, making the movement known in dressage as passage. He couldn't get quite the same height, relative to body size, as a grand prix horse, perhaps because the footing was soft. All the same, this wasn't a trot for going somewhere: it was a trot for looking at. For display purposes only. The distant deer watched, assessed and came to a decision, taking off at a fair old clip; the first deer broke into canter and made a token chase of 100 yards.

So I took my seat, well wrapped up because it was getting on for dusk and there was a frost coming. An empty mind is supposed to be the goal of meditation: my mind emptied all right, but it filled with thoughts of the sad plight of a friend's daughter. I sat on, caring but useless.

I picked out two more deer, both a mile off. In this place, at this time, darkness rises rather than falls: from deep grey hollows pheasants called and called.

Poor girl.

> Cold-weather gear, like religion and diet, is one of those personal frontiers, but here is my solution for the colder months of this project: on top of indoor clothes a lightweight fleece, a down waistcoat, a thick fleece, waterproof top and trousers. Head, hand and foot-gear dependent on circumstances.

Tuesday 24th

The cloudscape was a reckless assembly of blues and the sun between my shoulder-blades was not far from being warm. I could hear a bird calling in the tangles to my left, where the deer had laid up a few days back. This brisk churring continued for a few seconds while I puzzled.

Then I realized why I had failed to recognize it: no one was answering. It was one of the calls one long-tailed tit make to another as they forage, but this one was alone, and it flew off alone. One horse is no horse, as we horse-people say, recognizing the horse as a herd animal. One long-taily is no long-taily. Is solitude a terminal condition for a long-tailed tit?

And then a further moment of doubt: a bird, a harrier for sure, and the sun caught its wings for a second and for that second I was convinced: but then it was gone and there was time to doubt again. But I think it was an adult male marsh harrier, with the sun bouncing off those pale patches on the upper wing. But maybe it was just the brightness of the day.

For infallibility, dear reader, you must look elsewhere.

> The main foraging call of long-tailed tits is a convivial *si-si-si*. This second call – roughly *syrrrup* – is more to do with alarm or excitement. Information from the nine-volume *Birds of the Western Palearctic* (or *BWP*); an abbreviated form is available as an excellent phone app. This also contains recordings of birdsong, so it's doubly excellent.

Wednesday 25th

The sun was out, but it was cold and windy. I might have called it a miserable day had I not vowed, for the purpose of this project, to rise above such concepts. There had been a heavy frost overnight. My working day had been marked by one of those setbacks you are supposed to rise above.

But there was a heron sweeping down in an arc, as herons love to do, coming round to make his landing facing into the wind, always the best plan for a heavy aircraft like this. As he touched down he disappeared instantly. I stood up to look for him, marking the piece of vegetation that seemed to have swallowed him, had the bins on it at once – and I was looking at a deer, looking charming and faintly comic. That teddy-bear face of Chinese water deer can make them seem undignified beasts at times. The wild world throws these occasional comedic gifts at those who look.

There was a continuing passage of gulls from left to right, in the general direction of the Flood, mostly black-headed. It was then I noticed a mild itch on my face. Midge! I had been midged in late November! The brief hatch-out, the search for a mate, the strategy of using the cold time of the year with few predators about: it's clever stuff: but as I looked at the clearing sky I knew there'd be a frost that night. Would my assailant survive long enough to lay her eggs, those for which her blood meal had been essential? I did my best to wish her well.

> 🍃 Only the females bite; it's the same with mosquitoes. It's essential to the egg-laying process. It's not the blood loss that's the problem for us humans – they only take .001 ml – but the anticoagulant they inject with the bite causes itching.

Thursday 26th

I could feel the coming frost in the sun's warmth: not a contradiction because it would soon be night and beneath the clear sky, the world, unprotected by clouds, would freeze again. The sun, shining behind my right shoulder, lit up a deer 200 yards in front of me. As he – note the large canines – retreated into cover I could see only his two ears, lit by the spotlight sun. They glowed back at me like a signal.

I then heard a *thwock* from above and behind me in the ash. I turned and looked, but there was no seeing, and no telling if this was male or female. But I knew the species from the single sound, so there's a bit of omniscience for you: great spotted woodpecker.

As I settled comfortably into my space – perhaps I should say into my time – the bird carried on foraging. *Thwock*. Pause. Then another *thwock*. Sometimes two or three. Once there was a right-said-Fred sound of – so it seemed – falling masonry: a bit of excavation had gone particularly well. Beetle larvae live beneath the bark, and they were being dug out by the bird's power-tool beak. Perhaps I should put a warning notice at the foot of the tree: Bird At Work.

I sat for a while longer, reluctant to disturb the worker

above me. When I eventually moved, he carried on working without losing a beat: totally absorbed in the job at hand. At beak rather. *Thwock*.

> If a great spotted woodpecker has a red nape, he's male; if not, she's female. Juveniles have a red cap.

Monday 30th

My mind was full of geese. The pink-footed geese had been flying over the house morning and evening, sometimes in flocks of 2,000 or so, though the sight had been too thrilling for counting. But as I took my seat I had instead a flypast of 400 corvids.

Crows are the most intelligent of all birds, if we define intelligence to mean thinking like us. They are problem-solvers and tool-users. So why don't they fly in chevrons like the geese? Can't they work out that this would be more energy-efficient? Are they too independent-minded for such conformity? Or are their journeys too short to make it worth bothering with?

It was cold, dark, cloudy and I had forgotten my cushion. But I picked out an almost infinitely distant bird of prey: buzzard, harrier, sparrowhawk, I couldn't tell. It was beautiful even though it was the size of a grain of pepper. Was it beautiful because of what I knew, or was the speck beautiful in itself?

I wondered if a haiku master could capture this question – or better still the answer – in seventeen syllables. Ha! The narrow seat to the deep north, I thought, and allowed myself a smile. The deep northeast, anyway.

Then I stirred in my narrow seat, because here was something I hadn't seen from it before: a murmuration of starlings, or rather two of them. They were small as murmurations go, about 500 in each flock, oozing and shape-shifting – and then gone. In sight for no more than a couple of seconds. How come they never bump into each other?

I sat a little longer. If I was to reach a higher plane of consciousness that afternoon, I needed my cushion.

> You can find a starling murmuration near you by checking the RSPB website. Late November and December is the best time.

DECEMBER

Tuesday 1st

The meaning matters. There's more beauty in 'love' than 'shove'. The sound of creaking wood is not beautiful in itself, but when it means that your beloved is climbing the stair to join you, it's the most beautiful sound in the world. And though the chinking of a lone blackbird off to my left wasn't beautiful when compared to the song he would be singing in a few months, there was beauty in the thought of blackbirds all around, waiting for their cue to sing.

There were plenty of them about, mostly cocks. Many of them had migrated from Scandinavia to enjoy our softer winters. The females travel into southern Europe, where the chances of survival are better – but the males need to get back home as early as possible, to establish a territory, claimed and maintained by the beauty of song, before the females' return. So they spend the cold months with us, racing each other back as soon as the weather turns.

A young heron rose up from the marsh in front of me and then swerved sharply to my left, his right, to avoid me. Such a big wing. It seemed ridiculous that a bird with so big a wing can make a living in this crowded country.

The pink-footed geese had been daily executing their sky-filling calligraphy, with lines and their chevrons; I sat there wondering if they ever made real Japanese characters and wrote real haiku above the marshes of England. Then a marsh harrier got up and performed a manoeuvre like a wide but slightly compressed S. For a moment I seemed to see my initial burnt into the sky as if by a sparkler on Bonfire Night.

> Adult male blackbirds have the familiar glossy all-black feathers with the banana-yellow beak and eye-ring. The females are a quiet brown.

Thursday 3rd

It was really quite cold, with rain coming in great gusts. The sound of buzzards filled the air. There were three of them, four of them – no, one was a marsh harrier, longer in the tail, slimmer in the wing. There seemed to be some kind of discussion in progress: a significant piece of demarcation: who owns which piece of air at this precise moment. No violence: all was to be settled by the eloquence of those eight wings. The harrier slipped lower, to the level where harriers do their best work.

The three buzzards came together, mingled, almost merged – and then there were two of them, quite distant now, moving together in the same spiral, another form of aerial calligraphy, at home in the wind's buffeting embrace.

I heard the call of a curlew, and then the answering call of another – longer, sweeter – as if imploring me to stay where I was. So for a while I did, to the frank dismay, I fancied, of the weather.

> You can find a selection of curlew calls by searching for 'RSPB curlew'. In the most characteristic call they say their own name, but not quite as we pronounce it: a long stress on the second syllable, sometimes almost unbearably lovely.

Friday 4th

Nature has no duty to please me. Snow had fallen that morning for a good hour and there was still plenty on the ground as I stepped out in seven layers of clothing, feeling cheerful and venturesome. A magpie crossed the marsh in its swoopy way, remaining true to its nature despite the changes all around.

The wind from the southwest tested the layers around

my shoulders. Just below the horizon a tractor was towing a shooting-break: a covered trailer that takes the shooters to their positions. I once rode in one on some non-shooting occasion: it was like being on a laboratory shaker. How does anyone shoot straight afterwards? My host told me: 'Very few of them can – even before.'

The wind had stripped most of the remaining leaves from the oaks over the last couple of days. Two weeks from the solstice, the landscape was full of skeletons.

I had been brave enough to sit here: where was the spectacular sighting I was surely owed? But I was wiser now, knowing that the true spectacle was the cold, the silence, the whiteness of landscape, the faint sense of desperation, for this was a place and a time where only the best and the luckiest survive. And there would be worse to come.

A blackbird alarm call rattled off to my right; a crow rode the wind with the confidence of his kind; for half a second I saw the silhouette of a marsh harrier.

> A cotton scarf helps to keep your neck warm and you can use the protruding end to polish the lenses of your bins. My new fingerless gloves exposed only the top joint of each digit, enough to allow me to keep the lenses shining.

Monday 7th

No puddle on my seat: instead a piece of ice about the size and thickness of a dinner plate. There had been a hard frost that night and there would be another tonight. The air was still;

the cloud was low, but it was too cold for its insulation to make much difference. A great spotted woodpecker called from the ash, twice, three times.

I could hear the distant sound of the digger from the Internal Drainage Board, clearing the dykes. Soon it would be working the land in front of me. It's an invasive business that leaves the place looking devastated. They do the job, should the landowner agree, in order to keep the land drained, but the work creates dykes full of open water, good for wildlife. Certainly the herons approve, as do dragonflies and water voles.

A common gull mewed from the direction of the river. I was going through the going-through stage so smoothly I hardly noticed: what mattered were the lives and deaths in front of me. Many of the creatures that lived here faced a harder night than I had ever lived through. I sat listening, savouring each one of the seven layers. Layers of clothing, not layers of consciousness. If you seek enlightenment on a Norfolk marsh in December, you need plenty of clothes.

> Learning the call of the great spotted woodpecker changed the British landscape for me. Get that call in your head – *chick! chick!* – and you start to find these birds almost everywhere you find mature trees. Try the Woodland Trust website for more examples.

Thursday 10th

The digger had been hard at it all day, followed by a handful of gulls, three herons and a little egret. Now I was out there contemplating the newly exposed mud on a mild afternoon, the sound of the digger away to my left on the adjoining common.

And then I realized I was in the company of a barn owl – a female, I guessed, from the large size of her. She was working from a low flight, like a harrier, but with a quite different rhythm: urgent, hurried, always busy; when she went down she did so with a face-first dive.

But as she tipped forward, her head down, her long pale legs shot forward either side of her head: two eyes, two ears and two legs all operating in a line. I was reminded of a cricketer lining up a high catch.

Down she went: and stayed down. That meant success, at least for the owl. And then she was up and flying towards me, something big in her beak, perhaps a half-grown rat. I think she was planning to eat in the branches of the ash, but, seeing me, she veered to her right and vanished. Barn owls can't hunt in heavy rain: their flight-silencing feathers get clogged up. It's possible she had been close to starving: the rat was quite a prize.

About ten minutes later she was back, sitting on one of the benches, using it as a hunting perch. You can't possibly be hungry after that enormous rat, I thought. This use of a perch is a less committed form of hunting, but it worked all right. After twenty minutes of stationary staring she was down again, eventually emerging with another meal.

She went back to her bench; she seemed to have a short-tailed field vole, from its dumpy shape. She took it all down at once in a series of convulsive swallows. The success of an apex predator indicates that the ecosystem is in fair shape. No doubt the rat and the vole felt suitably flattered.

> 🌿 I learned this lesson years ago: if you come across something good, stay with it for as long as it lasts. I no longer think there might be something even better round the next corner.

Friday 11th

We had just bought a new telly, one with its own agenda and no interest at all in ours. No doubt we would master it eventually.

There had been plenty of rain in the last twenty-four hours, bad news for the barn owl, but now I was sitting in a not-quite break: the rain was falling with reluctance, doing all it could to hang in the air. The dome of pearly cloud was back. A heron flew over the marsh, and I remembered Kipling's adjutant stork in *The Second Jungle Book* flying 'as though each slow stroke would be his last'.

The digger's work was plainly visible; the heron started to work the exposed mud for protein. I felt as if we were both underwater: peering at each other through the depths like Jacques Cousteau and a shark. The light felt quite unlike the light you get on land: well, this part of the world has never been wholly solid, shifting, changing, rising and sinking. A few miles away there's a quaking fen that moves beneath your feet like a waterbed.

A jay flew over the marsh: it took me a moment to be quite certain of that, for it seemed drained of all colour. As did everything else. It was like trying to adjust the colour on the new telly. David Attenborough's face turned orange-scarlet, as if he had been on a monumental binge; we adjusted, overcompensated and reduced him to corpse-white pallor.

Now it was as if I had mismanaged the colour on the marsh: nothing but greeny-browns and browny-greens. I'd never realized before that this is what happens at the outset of winter: shorter days and colder days, days with the colour turned right down.

> I don't know if Kipling had a stern moral purpose when he wrote *The Jungle Book* and *The Second Jungle Book*, but if he did, he was far too good a writer to stick to it. Both books are full of wildness and I have been reading them off and on since childhood.

Sunday 13th

Occasional fierce gusts caught me in the back as I inspected the work of the digger. A blackbird was looking over it as well: not as good as when the earth was first shifted, but well worth trying. This was a young male: black plumage but a plain brown beak. A yellow beak marks you out as a rival to any other mature yellow-billed male: the callow, beardless appearance of the youngsters means they are less likely to get beaten up. That gives them a better chance of reaching yellow-billed maturity and breeding.

There was a flight of a dozen pinkfeet: nothing compared to the massive formations I had been seeing of late, but it was nice to record them from my seat and to hear that double-noted, slightly yodelling honk.

Many places further north, and also many at the same latitude but without our Gulf Stream coast, are almost birdless in the winter: the resources of the frozen land so scanty that anything that can fly away flies and anything that can sleep sleeps. This island loses many birds which go south in search of warmer weather, but others join us from further north. The pinkfeet come from Spitsbergen, Iceland and Greenland; many of the blackbirds, as we have seen, from Scandinavia. They have come to enjoy our winter: they remind us to do the same.

> Britain, being an island, has a lot of coast and therefore a lot of estuaries. Birds flock in from further north to spend the winter on great glorious gloopy British mud: easy sites to visit include Dee Estuary, Cheshire; Snettisham, Norfolk; Pagham Harbour, West Sussex; and Dawlish Warren, Devon.

Monday 14th

By afternoon the wind had dropped and the rain had stopped and there was even a little blue. Time for a sit: daylight time is tight when you're close to the solstice; I had missed out on more days than I wished in the last fortnight as a rather complicated book project reached its final frenzies of corrections and captions.

The high yowl of a buzzard was followed almost at once by a sort of grunting shout; this from a heron, now in sight making a big gliding half-circle round the marsh. Was something going on? I couldn't see the buzzard; the heron came down into the marsh, long landing-gear extended, and disappeared in the reeds.

About fifteen minutes later another or the same buzzard cruised into sight, looking heavy in the air. It took a perch in the big willow and tried to stare a meal into existence, hunting from a perch like the barn owl the other day. Six blackbirds, all cocks, flew over in a tight but untidy formation.

The buzzard sat on; perhaps it was going through the going-through phase. Then a brief fluster of wings as it shifted into one of the outside twigs, keeping nerve and his balance although the twig bent alarmingly under its weight. It had something in sight, and leant right forward, eyes and beak as far as possible from his feet, eagerness in the very line of its body.

The heron rose up and flew its weary way across the marsh – and then sighted the buzzard. At once it completely changed shape: from the classic heron's flight position of trailing legs and tucked-in neck, it seemed to stand up straight in the air, long legs a-dangle, neck outstretched, beak turned towards the buzzard. From there it let out two or three harsh calls: you'd have sworn it was swearing. After a moment, both birds flew off in opposite directions.

> 🍃 I think this was a stand-off between two big fierce birds, either of which could harm the other: you don't mess with me and I won't mess with you, okay? It's always good to ask the next question, the one that follows successful ID: just what the hell is going on here?

Wednesday 16th

There had been a shoot that morning on the far side of the dyke: a dozen figures dressed in browny-green and greeny-brown, all going bang while the fat balls of pheasants fell out of the sky.

I went out an hour or so after they'd finished, into the slightly eerie quiet. It was mostly that the pheasants had shut up; a world of fewer and quieter pheasants. But everything else seemed to have sunk into silence: all I could hear was the occasional *quop* of blackbird.

Then a bird was moving fast to my right: pigeon with the wind in its tail? Sometimes it all becomes clear in a single instant: the kestrel asserted his identity as a falcon and revealed his menace. It was something to do with the sharpness of those wings, slicing up the wind as he passed. It was something about the turn, the economical mastery of it all. How shall I put it? 'The achieve of, the mastery of the thing.'

They happen all the time, these little shifts of body language that reveal the species – and with it, what the species has spent all those millennia evolving for. Here was speed, control, self-certainty. There's a subtle pleasure to be had in such moments of revelation – and then the bird had gone, leaving the poor wind in tatters.

> These moments of revelation usually come as the bird moves or gets closer. But sometimes they come like a moment of enlightenment: as you interpret scanty information in a suddenly meaningful way. The more you look, the more often they come: it's about filling your brain with birds.

Friday 18th

I could see clearly now. This was not enlightenment from long hours of sitting; it was a new pair of bins. Swarovski had kindly and unexpectedly lent me a pair of their latest model, so I set aside the pair I had been using for the past twenty years to try them out. They were seriously bloody good, giving me an image of almost uncanny brightness. It's all about gathering the light.

It was my mother's birthday, a date we always mark by putting up the Christmas decorations, for that's the way it always was in my childhood. There had been another shoot, on land a little farther north than the previous one. The wind was also playing a part in keeping things quiet. I kept focusing these fancy news bins on distant pigeons.

Then at once there were two harriers in the air, performing complex manoeuvres. They rode the wind as I have ridden a young horse on my good days: with dispassionate confidence, with a perfect relaxed intensity. Did these harriers feel the same subtle joy in the same subtle mastery of a skill?

One of the birds went down and stayed down, perhaps having met with success. The other completed a circuit and then glided away, passing a small flight of rooks. Rooks routinely mob a buzzard and ignore a harrier. They watched this one pass, as admiringly, as I fancifully thought, as people watched me schooling a young horse.

> 🍃 Ah yes, binoculars. This is very much a value-for-money business: money buys you a bright and vivid three-dimensional image. It's not about magnification; it's about clarity: I could see better with these 8.5s than my old 10s. What to choose? There are plenty of helpful reviews online. It's also a good idea to try them out for yourself: there are shops at many of the big nature reserves. I was now using a pair of Swarovski EL 8.5x42.

Saturday 19th

It was warm, the temperature in double figures with a gusty wind. I sat down expecting at any moment a sudden shout of unseasonal song from a wren, and a minute later it came. I could almost imagine the singer apologizing for his incontinence – but it was good to have his confidence that spring would, indeed, come when the time was right.

It was especially good to have it now, as the second wave of Covid washed across the nation and the world. Well done, everyone! We've all done something deeply and desperately hard. Now as a reward, we must do something still harder. This is a cruel business, and no one is untouched.

But at least I could sit here and look and listen and think and not think. I had never reckoned herons as birds with much flying skill until I started this project. Here was a perfect demonstration of the art of the glide: a slow, controlled descent into the wind and then, with a subtle alteration in the shape and angle of the wings, the bird rose 20 feet in a single second, all without the hint of a flap. From this new height it descended at a leisurely angle and vanished into the reeds.

A second heron followed a couple of minutes later, touched down and remained in view: pacing and stopping, pacing and stopping, leaning so far forward in that classic heron attitude that you'd think he'd topple beak-first into the mud.

The sun was now low and bright and directly behind me, for it was very close to the solstice. A buzzard announced itself with a merry yowl, flew just in front of me at 50 feet – and exploded. He shifted, turned, revealed his underside and caught the sun like a great burst of fire, lighting him up in flames of russet and tawny set off with ashy grey. It was as if in that instant he had turned into a phoenix.

> Wrens are astonishingly loud for such tiny birds. If you hear a great shout of song from about knee-height, it's probably a wren; listen for the exaggerated trill at the end. They sing in spring and early summer but are prone to these mad moments of optimism on nice winter days.

Sunday 20th

It was a day of complicated weather: chilly, breezy, shifting patterns of cloud, intermittent sun. I paid this little mind: I was too busy playing the fool with the new binoculars, trying to focus on my feet. I could do it standing, but not sitting.

I knew it was a deer before I raised the bins and made the required hefty adjustment to the focusing wheel. And then I found another. They were both standing still: a staring match was going on. These teddy-faced Chinese water deer were exchanging the Paddington Bear Hard Stare.

Eventually one moved off, breaking straight into canter and then with a jerk dropping into a trot: *I'm not afraid of you, but it just so happens that I have urgent business at the other end of the meadow.*

The sun brought a little colour back into the landscape, and that gave me a clear view of a harrier, turning on the wind like yesterday's buzzard, though less dramatically, showing me the complicated underside of the wings. Later consultation with a few field guides revealed him as a third-winter male: soon to transition into the tri-coloured glory of an adult.

Then the cloud came back and the colour drained out of the world like water from a bath.

> Close focus is a good thing in bins: it's infuriating to find yourself walking backwards, away from a small bird, in order to see it properly. It's also good for butterflies and dragonflies. You can get ultra-close focusing bins, which are great for invertebrates: I use Pentax Papilio, surprisingly cheap.

Monday 21st

The earth beneath me was tilting back towards the sun. The solstice had been achieved at two minutes past ten, about four hours earlier. Now we were heading towards summer, though not in any itchin' hurry.

I was feeling thoroughly out of sorts. It had been a stressful day of work at the computer, but the real problem was the darkness. It gets to me every year: and it's like a visit from Harry Potter's Dementors. But I found a strange comfort in the wet weather: a light rain but each individual drop was unusually heavy, landing with a spatter on my hat and my waterproof trousers. Outdoor clothing, they call it. Perhaps it's really open-door clothing, or door-opening clothing: certainly it creates opportunities to explore places and times that would otherwise be forever unknown. And at least I was out there in what light there was: my clothing itself acting as a defence against the Dark Arts.

The thought gave me quiet pleasure. I was, after all, now able to savour the rain itself without conscious effort. Blackbirds were quopping sociably to each other, a handful behind me feeding on fallen apples in the garden; one or two checking out the digger trails on the marsh. A few goldfinches flew over with a jangle of merry notes, briefly perching in the willow to my left and then restlessly moving on, too early to roost just yet.

The shortest day. It felt like the shortest day ever. A wood pigeon roared over my head – that's what the wings sounded like, up close, not a sound I had consciously heard before.

Come. I still had chores to do. Half reluctantly I got up to do them, a brief curtain of drops falling from the brim of my hat, my mood, yes, a shade lighter.

> I have a fanciful mnemonic for goldfinch call and song: like someone scattering handfuls of gold coins.

Wednesday 23rd

It was dark like the end of the world. It was two o'clock and raining a little, but it felt at least two hours later. This wasn't an intellectual observation; it came from the gut: there had been a dizzying moment when I really thought I had lost two hours, perhaps distracted by the joys of correcting the corrections to a complex manuscript.

And then the news. We were back in another lockdown. My niece, who always comes to us for Christmas, was stuck in London alone, Cindy frantically organizing a last-minute hamper to comfort her solitude. And we wouldn't be making our usual post-Christmas trip to London and family jollities with my father. He was by no means demented, thank God, but was still having trouble getting his head round the idea of a pandemic. *Yes, I know, I understand – but why aren't you coming here after Christmas?* Like millions of others, I felt the pandemic as a personal assault on the foundations of happiness.

I could hear herring gulls, rooks, jackdaws. Very little else was stirring: the weather had disturbed us all. No one had any business out in this stuff, not even a naturalist. I thought of my friend Al in Australia, just back from buying the traditional Christmas dinner of lobsters and prawns, 'which we don't even put on the barbie at this time of year'.

Two swans flew heavily along the river, the only bright things left in the world.

When I turned my head I could just see a corner of the house, the Christmas lights casting a jolly glow across the garden. They had of course gone up on the evening of my mother's birthday, all ready for my non-arriving niece. I would be in there soon enough, warm and safe, ready to cook a meal rich with cheese and garlic.

But out on the marsh, out across the wild world, everything that lived was suffering: and it would be holding on tight through the darkness long after I had gone inside. Some of them would die before I sat here again, for that's how life works. I knew that already, but I had never before understood the principle with my gut. So I sat out a while longer before heading back to the merry lights and the still merrier warmth.

> It's all in Darwin, of course. *On the Origin of Species* was written as a work of popular science and remains thoroughly readable.

Sunday 27th

This project was supposed to record subtle transitions, the ones that pass almost unnoticed from one day to the next. But today I was looking at a world transformed: almost unrecognizable from the one I had gazed on before Christmas, for the darkness had brought with it thirty-six hours of intense and implacable rain.

Now the world was flooded: standing water everywhere. The embanked walls of the river had stood firm but the green meadows before it were now silver, mirror-bright under the cheerful sun. Two swans swam across a drowned field, behind them 100 black-headed gulls sat cosily on the icy water.

It was unspeakably lovely, though I wished I had worn my Christmas thermal socks. A female marsh harrier crossed, her pale head apparently in flames under the sun, like the Marvel character the Ghost Rider, who had featured in a Christmas gift to Eddie. The young male harrier followed a little later, the complex colours of his underside glowing like embers.

As if such spectacle was not enough, a great tit burst into song. It didn't sound like a serious attempt to set up territory, not yet, just an overwhelming feeling that such a thing would be possible in a matter of weeks. Winter officially began with the solstice a few days earlier: here was a subversive hint of spring.

On the far side of the river a tight formation of ducks rose and performed a rapid, low spiral in well-organized unison before dropping down again. At once another, different group did exactly the same: wigeon and teal, each in their own flock, appreciating a flood as only a duck can.

> A great tit sings one of the simplest songs you will ever hear: *teacher, teacher, teacher!* Sometimes described as a leaky pump. Great tits have many variations, but that's their default song.

Monday 28th

How absurdly gratifying. I sat down on a cold bright day, the floods still high, checked at once the harrier perch: and there she was, like an intimate friend.

There were swans to my left, swans to my right, finding all this water an excellent addition to their lives. I had been given a canoe for Christmas: not a solo craft like my kayak, but open, chunky and capable of carrying three. Earlier that day we had taken it around the dykes that form the boundary of the marsh and the common: they had never before been deep enough for such an adventure. We covered more than a mile, Eddie and me paddling, Cindy in the bows cutting us through with a pair of loppers.

It was a glorious voyage. We had taken a boat across the land: but in the Broads the demarcation between land and water is both fuzzy and volatile. Now, looking over the place from my seat, I felt a new kind of intimacy. I had walked it a million times; now I had paddled it. I knew it wet, I knew it dry, I knew it in all the in-between stages as well. I felt strangely enriched as a result.

Then, high above, I found a small skein of pink-footed geese: eighty of them, almost invisible to the naked eye, calligraphing their shifting lines at about 500 feet. The previous evening I had found a haiku by the nineteenth-century master Buson that echoed my fantasy of geese writing perfect Japanese characters across the sky:

> *Under a passage of wild geese*
> *over the foothills*
> *a moon is signed*

> Aimless scanning of empty sky can be profitable. You're looking for shape and movement – anything that breaks the pattern.

Tuesday 29th

Above the watery landscape all was calm, all was still: high cloud, a fair amount of blue. A group of twenty jackdaws flew over, followed by thirty more. I heard a whisper of goldfinch – on my left? On my right?

The temporary lagoon ahead and right showed me what the land would be like if they ever let the river reunite with its floodplain: a thrilling and alarming prospect, for the house we live on stands just as the valley begins its cautious rise above the level of the river. Then all at once a blizzard rose upwards from the waters: 200 black-headed gulls in the air together.

As if rejecting the notion of unity implied by the word flock, they at once became 200 individuals, each with a different agenda, flying off in many different directions or settling back down again. From behind the heronry I could hear greylag geese.

There was a mounting sound, fizzing and buzzing from the fallen willow ahead of me: a goldfinch with something urgent to communicate. After a moment's throat-clearing another goldfinch responded, this one behind me. It did so with a full and committed song, answering the tentative calls from the willow with the sound of spring from the ash. For a moment I sat there stupid and open-mouthed.

The willow bird joined in with a few pithy comments, which stimulated the ash bird into something not far from his full repertoire. In this moment of joy I felt the promise of frost in the air, and so, I had no doubt, could the singer above me. Long way to go.

> 🍃 There seems to me to be a continuum between goldfinch contact calls – I'm here, where are you? – and the full song of a territorial male. It's not always clear where one stops and the other begins. Is this the point at which the question of language crops up? Listen for those tinkly, jangly, coin-dropping sounds.

JANUARY

Friday 1st

The frost had been hard, deep and extravagantly white, but important family events in the course of our isolated Christmas had to take priority over everything else. I had missed a few days at the garden's end and I felt a little embarrassed by these gaps as I took my place. Yesterday there were 500 starlings rising and falling in the brittle air; I wondered if I would see them again now I could give them my full attention.

It was still cold, though not like the day before. January 1, feast of beginnings ... but my resolution had already been made back in September: sit on, sit on, and see what happens next. Two marsh harriers crossed paths near the river; at once fifty teal got up and performed a couple of circuits.

A heron landed on the frozen marsh and froze into the classic hunting pose of its kind, invisible to all prey in its motionlessness. Half a dozen birds flew over, well distant; I raised the bins rather idly and found a treat: lapwings, declining birds, birds I have seen in past winters, strobing past black-white black-white on floppy wings.

The water in the dyke before me was visibly lower, the casual lagoons smaller. The boat moored on the distant river was showing less of its superstructure. And then another treat: harrier, not quite over my head, flying north with easy purpose, a hint of sun lighting both the bird and the vegetation beneath her in a quiet harmony of reddy-browns – and, almost like a physical blow, I felt the most devastating and quite unexpected feeling of connectedness. It felt so powerful and private that writing it down seems like a breach of confidence.

> Lapwings, usually in loose flocks, are very distinctive in flight: broad, round wings that flicker in alternations of black and white. The springtime display flights are gloriously crazy.

Saturday 2nd

I sat down in early afternoon; it was cool, cloudy but bright. As I did so a heron 200 yards away jumped into effortful flight. Before I could apologize it reduced its payload by jettisoning all excess baggage. It must have shat a gallon.

I could make out three distant deer in front of the carr on the right, grazing unhurriedly, scarcely troubling to look out for wolves and tigers. A stock dove gave a hint of spring with three of its ardent coos. Birds were beginning to exercise their voices: they'd be needed soon enough and no one wants to be left behind.

A low flash over the marsh: possible sparrowhawk. As I was trying to make up my mind how possible this was, a kestrel flew over, higher and in the opposite direction. A group of twenty mallards, all male, crossed the marsh and then crossed back again.

I thought of the great science fiction writer Kurt Vonnegut, and the invented religion of Bokononism in *Cat's Cradle*. 'Busy, busy, busy, is what we Bokononists whisper whenever we think of how complicated and unpredictable the machinery of life really is.'

> Stock dove may be the most overlooked bird in Britain: observe a pigeon with green neck-patches and no white on the wings: listen for that rather saucy *oooh-uh, oooh-uh*.

Monday 4th

Clearly dissatisfied with its attempt at drowning all of Norfolk, the rain was back for another go. But it made a courteous pause as I stepped out, the wind blowing spiritedly from the northeast, straight into my face. I remembered our Christmas paddling across this landscape or waterscape, looking up from the dykes like herons, the sky bigger than ever before.

And as I sat, the sky at once rearranged itself, dividing itself in two. To my left over the heronry it was an averagely bright winter's day, a little blue, a little sun. To my right over the second carr, the sky was unapologetically black, lit up in the eerie way that precedes a storm.

A rainbow appeared with the modesty of a star actor trying not to interrupt the show by provoking a too-deafening round of applause; it came to an end more or less on the spot where the three deer had been grazing the other day. A swan, standing on the riverbank, opened its wings in brief heraldic exuberance, glowing as if caught in a spotlight.

And then of course the rain, but by now I was in great good heart and the weather could be savoured, or at least experienced for what it was. A marsh harrier flew from the dark side into the light, almost self-consciously symbolic.

I sat on, thinking about bananas. I was writing a chapter about bananas in *The History of the World in 100 Plants* and had just learned about the banana apocalypse: the loss from our lives of the tastiest banana ever cultivated, the Gros Michel. Basho, the great contemplator of nature, named himself after the banana plant that grew by his hut. For a while longer, I contemplated the idea of contemplation. You go outside to study nature and wonder if what you're really studying is yourself – but you soon get over it. That's just the going-through phase. Before long, you've marvelled enough at your marvellous self, aware that there are greater marvels available. The wind and the rain stepped up but I didn't mind.

> I had been given what's been called a snood for Christmas – inaccurate name (a traditional snood is a bag for putting your hair in) but a useful garment. It covered my face from just below my eyes and ears and made the face-first wind easily bearable. Note that I left my ears exposed. I'm a wildlifer: I need to hear.

Tuesday 5th

There was a distinct lack of courtesy from today's weather. I took my seat in cold, wind and rain. I wasn't expecting an immediate reward for my intrepidness, but I got one anyway: a kestrel hovering just this side of the river.

It held the position in the air for longer that you'd think physically possible, staring down at the ground with a fiercely angled neck – and then, with the drastic change of gait that kestrels go in for, it dipped a wing to catch the wind and went in an instant from stationary hover to flat-out charge with the northeast wind in its tail, making a shallow right-handed curve as it went out of sight, as a skater's heel sweeps smooth on a bow-bend, and reminding me once again that wings are even better than words.

I was still nodding approvingly when it came barrelling back, missing me by about ten yards, an early supper, interrupting the sleek aerodynamic shape, dangling from its talons. Showing no problems from the additional drag, the kestrel vanished to the right-hand carr.

A marsh harrier passed in the distance; a heron announced its arrival with a loud call: this is my marsh, and what's more, my personal latrine. At this, 100 ducks took to the air, performed a circle perhaps ten feet from the ground and dropped back down. A little later I walked, feet sloshing in the standing water, around the marsh to check them out: teal and gadwall had taken up residence on this temporary lagoon left by the Christmas rains.

> You know it's a kestrel as soon as it starts to hover, but it's worth staying with the bird once it's started. Look at the uncannily still head that makes all this turmoil work as a hunting technique, and at the fanned and depressed tail that keeps the bird precisely in place. A kestrel faces the wind and flies at the exactly the same speed as the wind: that's what hovering means to a kes.

Wednesday 6th

It was a little milder, the air almost still, the snood back in my pocket. Before I had reached my seat I heard the sharp, repeated cries of an excited kestrel. A second later I saw two of them, one in hot pursuit of the other. I sat down, wondering what was going on. They started up again and I jerked my head sharply as they passed, the pursuer calling loudly, the pursued giving everything to pure speed.

Rules are, of course, meant for breaking, so I abandoned my seat and took a dozen paces left – just in time to see a single kestrel fly into the big oak in the meadow. He – clearly a male – was sitting perfectly lengthways on a convenient horizontal branch; as I watched he dipped his beak to his talons and pulled a long strip of flesh.

I had missed the crucial part of the story. Had he stolen this fine meal? Or had he saved it from his pursuer? I would never know, but since he was now alone, I had him for the thief. A kleptoparasite, to make him more respectable.

I returned to my seat. Searching the sky, I found a distant flock of 500 birds. No idea what they were, far too distant, no helpful flying formation to aid a little guesswork. Carl, a friend of mine and a crash-hot field naturalist, had recently seen four nenes in a flock of barnacle geese; nenes are the Hawaiian geese that Peter Scott saved from extinction at the beginning of the Environment Movement in the 1950s. For all I knew, this flock was full of them.

A flicker in my peripheral vision caused me to shift focus drastically: five yards away two robins were comparing the redness of their breasts. After a few moments they flew at each other and engaged in ritual combat, breast to breast, red to red. They separated and perched a couple of feet apart. One then gave a brief snatch of triumphal song; at this the second bird left. Defeated but undamaged: knowing when to back down is an important survival skill. Only equally matched robins fight to the point of desperation.

> 🍃 An adult male kestrel is easy to tell from a female: smaller, neater, more orange, and with a dapper grey head and tail. The females are barred and speckled. The second bird was too quick – and perhaps I was too caught up – for a diagnostic view.

Thursday 7th

When the air is very still you can hear the train. It runs on the far side of the main river, its nearest point close to the confluence, about four miles as the kestrel flies. You can reach it via the chain ferry; if it's closed or you're a heavy vehicle or you don't want to pay the toll you must travel 25 miles to reach the same point on the opposite bank: part of the logic of this watery landscape.

I could see nothing moving, which added to the mild eeriness of the day. Sounds came through with slightly unnerving clarity. There was a constant quopping from a couple of blackbirds, both of them feeling restless, the tiniest hint of spring.

A buzzard yowled, somewhere close but out of sight. I could hear the grumble of the road. A crow gave four loud, rather pedantic caws. The greylag geese, gabbling on the Flood a mile off, were clearly audible: a wigeon whistled from the flooded meadow.

And then with great swagger eighteen shelducks came flying in, white feathers, red beaks aflame in the clear light, chestnut stripes across eighteen breasts. They looked infinitely dashing as they made a small circle and then, as if called to order, formed a proper V-formation before powering off together in the direction of the Flood.

> 🍂 Ducks don't necessarily quack; the fact that some ducks whistle has come as a pleasant surprise to many a beginner. Drake wigeons whistle – the French call them *siffleurs* – and so do drake teals.

Friday 8th

I wish that my incompetence with computers saddened me. But it doesn't: it sends me into fits of black rage. Empirical evidence suggests that computers don't change their behaviour when shouted at, but it's a truth I am reluctant to embrace. Such episodes are invariably followed by a still blacker shame. At last I took my seat on a cold, still afternoon, overjoyed to step away from my own failings.

I didn't compose my mind. The birds did all the composing necessary: I could hear (busy busy busy) robin, long-tailed tit, goldfinch and blackbird. One of the blackbirds stopped quopping and started chinking: that seemed to me significant. The chinking is louder and more continuous: more social too, since it seems to seek and frequently gets an answer. It's a precursor to the spring song which comes much later: blackbirds chink together at dusk towards the end of the winter, in a strange compromise between rivalry and solidarity.

And then a wren sang. Actual song. It wasn't the full ringing verse that ends with a prolonged and almost violent trill, but it was a sincere attempt at the opening phrases. He did this again, altogether four chunks of uncompleted but quite genuine song.

A deer called from half a mile away: still pretty loud, giving it absolutely everything. I saw the head of a swan – the head and nothing else – moving smoothly across the countryside for about 100 yards before disappearing. Water in the dykes still pretty high, then.

> Both muntjac and Chinese water deer bark, and both sound like the Hound of the Baskervilles. I have sometimes heard the roars of muntjac in suburban woodlands at dusk; plenty of examples of both on YouTube.

Sunday 10th

My seven layers of clothing were not enough for a prolonged sit. The previous day Eddie and I had attempted one more paddle round the dykes again but had been forced back by ice; the paddles skidded off the surface unless you drove them in as if you were digging a trench. My fine new craft had failed to follow in the wake of Captain Scott's *Discovery*.

A kestrel flashed across the marsh, approached a tree as if determined to knock it down, stopped on a dime and in the same movement rose three feet to step composedly onto a branch like a commuter boarding the 8.19. He then fluffed up all his feathers for warmth, losing that sleek falconine profile, but as I looked through the bins I could see his head moving, his eyes hard at work, scanning the cold wet vegetation for a meal. Then he was gone, impatient, living larder empty of anything easy.

Behind me I could hear the triple-note of long-tailed tits: I half-turned and caught the movement of a foraging party in the fallen willow. There are two landscapes out there, one inhabited by me, one by everything else. I wanted to write about the irrepressible jauntiness of long-tailed tits, as if like boy scouts, they smiled and whistled under all difficulties.

How jaunty were they actually feeling? Days like this are brutal for everything that lives here, except me. The delicate pink beauty and the irrefragable perkiness of the long-tailed tits filled me with good cheer.

> Once I had got that *si-si-si* call of long-tailed tits lodged in my mind they miraculously changed status, going from mildly unusual to ubiquitous. If you learn that call you change your world.

Tuesday 12th

It was almost warm and the sun was throwing colours at the marsh in a reckless fashion. Even the grass at my feet was greener than usual. A distant harrier looked almost red as she turned away in that long, lazy curve that harriers do so well.

Far ahead, where the land rises unhurriedly to the lip of the shallow valley, a tractor was pulling a power harrow. It looked as if it was wearing a bridal train: 200 black-headed gulls followed close behind, dropping onto the newly turned earth for exposed invertebrates. The soil was in decent condition or they wouldn't have been there.

A great tit sang a few bars of song: day by day, birdsong was reconquering the land. A lone lapwing flew over: it should have been in a flock of several dozen.

When I make a journey to look for wildlife I am, I suppose, looking for drama: the day-maker: a sight to boast about. By sitting I have grown increasingly content with ordinariness, with quietness, with the unremarkable daily occurrences – and even as I sat there was a moment of drama.

Sparrowhawks are fierce and fast, and they're usually in sight for half a second as they seek to seize lunch out of the air. I saw her for every bit of that half-second, perhaps a little longer, her purpose and her menace briefly filling the marsh ... and then she was gone. In in time-honoured birder's fashion I wrote the world *sprawk* in my notebook. It seemed rather inadequate.

> The strong flap-flap-glide rhythm is a good clue for a sparrowhawk. The females are a third bigger than the males, and brown-backed where the males are slatey-grey. On a hunting run they will often fly so low you think their wingtips must brush the ground.

Wednesday 13th

It was cold, dark and dreary: the sort of day my mother used to see as a prevision of hell. They sky was busy with mallards, but the peregrine I had seen earlier that day was long gone; perhaps she had only turned up to teach the sparrowhawk what drama really means. But now it was just cold and determinedly undramatic. Two blue tits in the fallen willow were showing a mild interest in each other. This wasn't really the stuff of wildlife documentaries: the rapid succession of hunts, deaths, fights and courtship. It was more like a screensaver.

Then it started to rain. No, it didn't. It started to hail: see the stones bounce. Gang, gang, the hail's all here. Raining pearls. That poetic thought didn't make them any warmer. Or softer.

And then without any hint of drama whatsoever a harrier flew across the entire width of the marsh, a long glide with her wings held high, interrupted by a few easy flaps: and suddenly I understood why I was here, why I was sitting out in the hail, why I had been sitting here on and off for sixteen weeks, why I would sit here for another thirty-six, why I would continue to sit here long after this project was done.

I tried to make sense of this moment of perfect confirmation. The nearest I got was the *Goldberg Variations*, a piece of music I love above all others, and the 32nd movement, the last, when the aria, the perfect melody that begun it all, returns, both the same as ever and more marvellous than ever before.

Unmelted beads of hail fell from me as I rose to leave.

> Peregrines and kestrels are both falcons, and make a similar sharp-winged silhouette. It's easy to mistake a kestrel for a peregrine; almost impossible to mistake a peregrine for a kestrel: the rule that works for hobbies works for peregrines.

Thursday 14th

It had rained most of the night and was still hard at it when I stepped out, a good solid rhythm punctuated by sudden gusts. You can get waterproof notebooks, but I've never tried one. No such thing as bad weather, only bad notebooks. Splat: my note about a lone quopping blackbird gone forever.

It was raining fiercely enough for individual drops to be visible in the bins as I scanned the land before the river, two harriers, one above the other, showing a wary tolerance of each other. This wasn't a day for extravagant confrontations.

On days like this I never rode a horse on the roads if I could help it: the rain blows into a horse's eyes and the wind rushes past the ears, so the most important senses are compromised. If a prey animal can't sense danger it becomes deeply uneasy: so my horse would dance and spook.

It was just as difficult for the creatures of the marsh: hard enough to hunt, harder still to avoid hunters in these times of sensory deprivation. On the meadow by the river I could see two swans hunkered down, faces protected from the driving rain.

> You, the human observer, are inhabiting, at least roughly, the same sensory world as the creatures you are observing. A well-judged imaginative leap will allow you to understand them a little better.

Saturday 16th

The colours of the marsh had changed again and changed drastically: the green, but not the brown, had been replaced by white. Snow covered the ground and an icy rain was falling. The air was full of noise from magpies and crows: something was up.

I thought I had it: there was a kestrel in the nearest willow, trying to penetrate the snow with its gaze and find a meal. But the cacophony wasn't centred on the kes. I was puzzled: but then a buzzard called and the mystery was solved. It was perched in the next willow along and had collected a ring of admirers: four magpies and four crows all in the same tree, the crows cawing and the magpies chacka-chacking.

The buzzard kept perfectly still. He had a plan: he was going to outlast them: you'll get bored before I will. I was reminded of something, couldn't put my finger on it – but yes, the wonderful Breugel painting *The Hunters in the Snow*, with crows and magpies in and around the skeletal trees in the frozen landscape.

Another buzzard appeared, making a beeline for the kestrel's tree, and the kestrel abandoned perch at once. The buzzard tried the place out for a minute and then flew on, dissatisfied. The first buzzard was still sitting tight; his attendants now reduced to two magpies. All the bounce had gone out of them: silent and still, they watched the unmoving buzzard and wondered what to do next.

And like the buzzard I looked out across the unspeakable beauty of the white marsh and the hunters in the snow.

> Birds are good birdwatchers, so it's a good idea to pay attention to them. Many a time the sounds and movements of an everyday bird have led me to the discovery of something special.

Sunday 17th

The thaw was almost complete: just a few patches of snow left in sheltered spots. A cool sun was shining and a great tit was welcoming its return. Did this count as song? It wasn't the default *teacher-teacher* sound, but great tits are notoriously variable. This was a more complex phrase of five notes.

It might have been a contact call, but it didn't feel like that as the bird repeated the phrase a dozen times. It sounded more like a tentative anticipation of spring: for you must sing of spring even when there's snow on the ground. I remembered another chunk from a Modesty Blaise thriller: two characters are wondering if Modesty will save them. 'If you're playing a hand of bridge and you can only make your contract if the adverse cards lie in a certain way, then you assume that's how they lie and you play accordingly.'

A robin sang; there must be some indiscernible moment in a robin's year when a winter song becomes a spring song. Then a stock dove tried out a few phrases: assuming that spring lay ahead, just out of sight. Steady on, chaps. I know it's stopped snowing, but let's not get carried away.

> Advice from Bill Oddie in his essential *Little Black Bird Book*: 'If you hear a call and don't recognise it – it's a great tit.'

Monday 18th

I seemed to be in a locked room. It had a roof of cloud that stretched before me for perhaps half a mile; beyond it a wall of solid floor-to-ceiling blue. Against this a harrier moved, invisible to the naked eye but found and then re-found in the miraculous bins. Marsh harriers set the world aright. Difficult morning? Try a harrier.

The lagoons on the distant meadows had almost gone. A buzzard called sharply: I found it just as a crow launched into it. The buzzard wasn't taking this calmly: it half-rolled in the air and, inverted, went at the crow talons-first. The crow avoided this without difficulty: a buzzard can out-soar almost anything, but the crow had the edge on agility, attacking him again and again, short darts from slightly beneath the buzzard: so agile it could ignore the old dog-fight rule that states that height is an invariable advantage. The crow didn't let up until the buzzard had retreated beyond the furthest carr.

A kestrel made a long, low pass across the marsh, hovered briefly and dropped. He came up empty-clawed, taking a perch on the lone alder, staring down. Almost at once he was down again, and this time he had caught something. He went back to the perch – and to my astonishment I saw that he had caught a snake. The damn thing was about a foot long and as thick as my thickest finger. Whatever impulse had made this snake break out from hibernation, it was a serious error of judgement. The kestrel ate it whole, head-first, in a series of convulsive swallows. An inch of tail protruded from the open bill as the bird paused for a last effort.

> 🍃 There's a lesson here about recording what you see rather than what you think you ought to see. I doubted my own eyes; in theory grass snakes – the only snakes I have seen on the marsh – are still in hibernation in January. Carl said yes, it's unusual, but he had occasionally seen snakes in winter. Well, I wouldn't be seeing that one again.

Tuesday 19th

The wind was coming in a series of powerful gusts that left the trees scrabbling at the air, especially the willows, with their thin whippy twigs. The cloud cover was total. A great spotted woodpecker flew over my head and carried on with the southwest wind in its tail: it wasn't in sight for long. I could hear the murmurs of birds but it was the sound of the wind that dominated. I sat for a while longer and then felt it was time for a marsh harrier.

And there she was again, gracious and obliging bird. What an exquisite turn: using the wind to bring herself round under complete control, just a shift of the wings, one higher than the other, to bank through 120 degrees. She then crossed the marsh in a long and slightly wobbly glide.

I could forgive the wobbles; she was flying at 90 degrees to a wind that was gusting at 40 mph. A bird can glide in a straight line through a crosswind, travelling faster than the wind: using the wind's power to defeat the wind.

Eddie and I had been throwing a Frisbee that day: it tended to jump in the wind unpredictably – but the harrier might have been cruising on a summer day of perfect stillness. Was she anticipating and pre-empting the irregularities of the wind? Or were her corrections so sharp I couldn't see them? And there was me on the ground wondering if the wind would blow me into the dyke.

> A sailing boat can travel much faster than the wind: that's because a sail is essentially a wing: it works on pressure difference, not shove. Boats in the Americas Cup can travel three and four times faster than the wind: there's a record of a New Zealand yacht hitting 50 mph in an 18-mph wind.

Wednesday 20th

The chair had blown over, but at least I wasn't in it at the time. I picked it up and placed its feet in the sockets that had formed over the past few months. As I did so I found a pencil; it still worked. Behind the sound of the wind I heard a jumble of goldfinch sounds: a dozen notes in search of a tune.

Four buzzards formed a lofty ring, revelling in the wind; thirty black-headed gulls made a brief spiral. A little later the marsh harrier made her customary appearance; at one point she rose 20 feet vertically in about half a second, with no discernible movement of her wings. What must it be like to do that? As if by a small shift of my body shape I could be sitting on a branch halfway up the ash.

The male kestrel flew across the marsh: I suppose another snake's out of the question. As he did so I made out a small movement on the ground by the long dyke that runs away from me, no doubt a dunnock or a wren, both lovers of low places. I turned the bins on it from the idlest curiosity – and it was a kingfisher. As I focused the bird flew up to a prominent perch above the dyke and commenced staring down at the water in all its improbable colours. I sometimes wish I could donate a few of my kingfisher sightings to non-birding friends who long to see one. But you have to go where kingfishers are, and it helps if once you're there you sit still for a while.

The kingfisher paused for moment and, perhaps feeling the perch was rather exposed, dropped into the shelter of the dyke and flew along its length, leaving a long blue streak on the retinas of my eyes.

> Look for a movement: a small dark shape flying close to the water. Kingfishers don't always look colourful; it depends on the light. Kingfishers in flight make a chunky, almost triangular shape: follow it with your eyes – they're often too quick to use your bins – and sometimes the bird will catch the light and make your day. Or year.

Thursday 21st

I was late that day, what with one thing and another, but I had a quiet hope that this might work to my advantage. And damn me, that's exactly what happened: even as I approached my seat I could see a barn owl, shining white through the grey, stooping low, regaining height, dropping to the ground and coming up empty.

Every nuance of body language expressed the idea of hurry: a very limited time for a very important task. Barn owls operate best in the No Man's Time between night and day when they can best exploit their sharp low-light vision and acute hearing. This one was going to search every square foot of marsh before dark.

I watched the owl as colour drained from the land, realizing that the bird's colours were exactly those of the marsh under snow, white and golden brown. Urgently, the bird scudded westward to see if the Common would provide a meal.

I stayed a little longer, savouring the luxury of the dark. A blackbird chinked. A wren sang half his song and then stopped short, as if embarrassed. A second blackbird joined in the chinking. Soon the sky would be as black as a blackbird.

> Birds that are active at dawn and dusk are crepuscular (fine word). Barn owls will hunt in the day when they have to, but they prefer low light to direct sunlight. They sometimes appear when the sky blackens before a sudden storm.

Friday 22nd

It was sunny, still and cold: in the marsh's beauty there was a firm promise of still colder weather to come. All the same, a great tit briefly tried out the *teacher-teacher* song, in a manner that seemed to me unhinged. A harrier flew over; she had lost a primary feather in her left wing. Didn't seem to bother her.

I had begun this project in secrecy, perhaps because it

seemed unhinged, perhaps because I didn't know if I'd stick to it. It was probably the best way to begin: but I had recently come clean. My family now knew what I was up to, and if they thought me unhinged they were too polite to say so, at least when I was listening. So it was easier to get on with it.

A wren was ticking to my left, down low, where wrens like to be: lovers of tangles and deep cover. I heard a faint fizz behind me, followed by more ticking, this time on my right. The wren had flown behind me, very close and no doubt at knee-height, to another tangle.

I could feel frost in the air like a great hand about to clutch.

> That sharp ticking is a wren's alarm call, a warning to other birds of a threatening presence. It's also information to the threat itself: you've been spotted. You can listen to this by searching 'wren alarm call': astonishingly loud for so small a bird.

Monday 25th

There was snow on the ground again; a light dusting that made a frozen patchwork in barn owl colours. It was sunny with a persistent wind. In the clear sky the not-quite-full moon looked a trifle embarrassed.

Four buzzards manoeuvred around each other above the marsh, in and out of each other's airspace. What constitutes social distancing? It was a question that had preoccupied us all for the past year: but the question of personal space is one that concerns many species: one that varies with circumstances and the time of year. Who was doing what to whom up there in the sky? Buzzards are notoriously variable and distinguishing the sexes takes a better eye than mine. This seemed to be a stately dance, but who was whose partner?

A crow attacked one of them but its heart wasn't in the job; too many buzzards to feel entirely safe. Two of the buzzards were getting more and more interested in each other. There was nothing ostentatious about this: all gentle circles and spirals and sweeps, a demonstration that a big heavy bird can be as light as – well, a feather. Was this the wings of love?

One of the other two birds perched in the willow near the river; after a while another flew to the same tree. The perching bird didn't take this lightly: both birds suddenly assumed fantastic heraldic positions: one that a herald would describe as 'wings displayed and elevated'. They looked like a pair of angels. Rather fierce angels: was this love or war? The other two buzzards were now locked into the same perpetual circle, slowly moving north until they disappeared behind the heronry.

> There's a line in Anthony Powell's *A Dance to the Music of Time* that seems to me to sum up such moments, for all that it's about social adventure rather than solitary wildlifing: 'Life is full of internal dramas, instantaneous and sensational, played out to an audience of one.'

Tuesday 26th

More snow, more frost. But I was up for it: and as I looked from the window of my hut before stepping out I saw a stoat on the bank of the dyke, bounce-pouncing along in his stoaty way.

But outside it was bloody desolate: a blasted heath or marsh. Corvids were calling, crow, rook, magpie. Who could believe that this awful place would be a feast of song within a few weeks? The seasonal lands are as miraculous as any rainforest.

A kestrel crossed the marsh at ten feet and then performed an extraordinary halt, spin and drop, flouncing his tail as a flamenco dancer flounces her dress. No good came of it: he left empty-taloned and sped off towards the Common.

There's a reason for our perpetual busyness, our incessant search for distraction. Sitting alone and quiet is sometimes the pleasantest thing, but not always. Every so often some secret part of you seizes the opportunity to go through a catalogue of the worst moments of your life. Why does it never take you through your greatest triumphs instead? Always so many more of the former, I suppose. In a few moments I went from easy content to face-hiding awfulness.

And then across the garden I heard a bird singing out with complete conviction, a bird that saw only a vista of hope.

Shamed by a great tit.

> A lot of wildlifing is about retraining your response to information from peripheral vision. The idea is to stop tuning out small distracting movements and instead turn towards them. That's how I saw the stoat.

Wednesday 27th

It was a nice mild day with high cloud. As I took my seat a sparrowhawk powered by in front of me, travelling low and fast, heading into the meadow and hoping to surprise a small bird along the fence-line.

The barn owl showed up again, making three busy, searching circuits of the marsh. She dropped twice, but caught nothing, or nothing more than a beak-full. The owls and the kestrels have been giving the marsh a hammering in the past fortnight.

There's a marvellous moment that comes with barn owls: the illusion of eye-contact. As they fly straight towards you, that round, flat, rather human face seems to respond to your own. I forgot I was looking through binoculars: it was as if the owl was looking straight into my eyes and making a dispassionate examination of my soul.

And then she was gone. A wood pigeon sang his full and much-worked phrases of song: give it all you've got, pigeon, because it'll be relegated to the background very soon. As I stood I heard a sudden cry of kestrel: had they been tangling with the barn owls again? Have the kestrels pinched another meal?

> Traditional mnemonic, delivered here with appropriate apologies to all Wales: wood pigeons sing advice to the cattle-rustling Welshman: steal *twoooo* cows, Taffy.

Thursday 28th

The rain had been heavy overnight and a little was still falling. But high in the ash tree the great tit was singing his five-note song again. No, six. Then back to five. About 20 yards further on, another great tit sang the same song back, answering the challenge. A boundary was being established.

A buzzard called; the harrier made her traditional flight over the marsh along the line of the river. A robin sang above me once the great tit had dropped silent, the better to get his message across. The male kestrel was perched in the nearest sallow; he shook himself like a dog to get rid of some of the wet.

The robin sang on. Another joined in a small distance away. The first robin responded with more and better song: I can be more plaintive than you, pal. Then a pied wagtail called: all around a sudden lurching prevision of spring's plenty. In the sallow the kestrel got down to work on aircraft maintenance, preening his flight feathers with all the confident care of a craftsman.

> Song is a challenge and must be answered in kind. But it's a way of avoiding violence rather than seeking it: song is usually enough to settle a dispute. It can sometimes escalate into display: a pose-off. Physical assault, as seen a few days earlier, is the last resort.

Friday 29th

A sunny morning had led to a windy, cloudy afternoon, but the robin was singing in the ash with some intent. Another was singing away to my left: another boundary. A movement near my feet: a third robin was working the tangle the wren had been exploring the other day. I turned towards him: he was interested in me, rather than alarmed, bouncing from one perch to another, giving me the once-over. Curiosity and boldness are part of a robin's strategy for life.

Above the distant valley wall a sudden cloud of corvids rose into sight, rooks and jackdaws, I assumed, though they were too far for accurate identification by either sight or sound. The flock rolled and spread, and then linked up again in a celebration of togetherness. There's a huge corvid roost along the big river beyond this little valley, and on big nights they gather there in thousands, though rarely visible from here. Here were a few hundred.

And just as suddenly as they had arisen, they were gone: their disappearance as inexplicable as their appearance.

> When there's wildlife close by, it's always a good plan to move in slow motion. Raise your binoculars as if you were doing tai chi and you'll be less likely to scare things away. The more rapid your movement, the more alarming you are.

FEBRUARY

Monday 1st

Lord, I was glad to be gazing out over the marsh on this cool and cloudy afternoon. I had spent much of the morning on the phone: my father thought he had made a massive financial cock-up that had left him destitute and his children penniless. He was in an awful state. He wasn't demented, not by any means: but he was prone to confusion and then by easy stages to panic, especially about money. Eventually we sorted it out, but the morning left us both shaken.

So there I was back on my seat looking at a female harrier on a different perch, one like a hitching rail in the middle of the reedbed. Another bird of prey flew over on long floppy wings revealing, an instant later, a forked tail. Red kite: not – or perhaps not yet – a daily bird here, though there's a growing local population. They're carrion feeders and in East Anglia we don't bother with tarmac: we pave our roads with dead pheasants. I watched the bird perform three easy circles of the right-hand carr before continuing southeast.

A great spotted woodpecker was drumming on the far side of the meadow: not a search for food but a spring song in the form of a drum solo. I could swear he was using the barn owl box for its resonant qualities.

Four swans came parachuting down into the river, wings still, great black feet stuck out and spread. Why do they do this? I guessed that they use them as an aircraft uses its flaps: the splayed feet creating an increased aerofoil surface, at the same time slowing them down and allowing them to fly safely at a lower speed. A pilot puts down flaps for landing; so does a swan. That's why a swan can land on the spot with a relatively slow landing speed, but needs a runway – like 25 yards of open water – to get into the air.

I felt happier for working this out. I was easier in my mind now. Not less troubled, but more capable of bearing my troubles.

> It's always interesting to make a note of such speculative fancies, and to check them out later. Turned out I was quite right about landing swans: most ducks and geese use the same technique but on a less obvious scale.

Tuesday 2nd

Seasons overlap. At this time of year a single day can shift in an instant from spring to winter and back again. Right now I seemed to be in both seasons at the same time: warm, cloudy and bright with a singing robin and a drumming great spotted woodpecker.

Earlier that day I had checked out the barn owl box, kindly set up by the Hawk and Owl Trust and most years sheltering a pair of jackdaws. It showed no signs of use as a percussion instrument and, besides, the bird was drumming away as I looked. He had found some dead branch on the far side of the tree, one that gave his drumming the resonance and carrying power he was looking for.

As I sat on I heard a call I had never heard before: *a-kiss-kiss-kiss*. Repeated. And then again. After a moment I realized this was Bill Oddie's Law in action: the timbre was very great tit. They love to be versatile: for them, an exotic call is as good a come-on as you could hope for.

A greatspot flew over the marsh, up-down, up-down: like an aerial showjumper. The bird flew on, climbing a little higher with each bound. More song, more song: how many robins? Three? Four? Impossible to be accurate without moving. Perhaps the only way you can count singing birds from a fixed point is one, two, many.

> 🍃 An awful lot of bird-census work is done by ear. The first skill for a census-taker is to learn birdsong; the second and harder skill is to count each singing bird only once.

Thursday 4th

The morning had been unambiguous spring; by the afternoon things had been reined back towards winter. It was still lovely enough: sky horizontally striped with mauve and purple clouds, a skyscape that would be sentimental in any medium but reality.

There was a southeast breeze in my right ear and the afternoon was full of song: robin, collared dove, wren, all hard at it in the no-time-to-lose manner of spring. A great tit offered a more conventional phrase than the one uttered on Tuesday.

For the first time in many weeks a stranger who came here by chance would not have thought me wholly mad. Or perhaps I was no longer aware of my own madness. Well, never mind: here was a song thrush, about 100 yards away. After a slightly tentative start he got his confidence up and started to sing properly: a series of ringing and repeated phrases.

We are always reluctant to acknowledge the good things in our lives for fear we will jinx them: celebrate too soon and the joy will be snatched away. Perhaps this mindset comes from the teasing advances and retreats of spring. But today at any rate, out here in the landscape of early February, spring could be given full acknowledgement without fear. It was even possible that tomorrow would be as kind and generous as the day that was drawing to a close.

> Song thrushes start to sing in this second phase of spring. They will take a nice phrase, repeat it two or three times, and then discard it for another: and once they get into their stride a single bird can fill a landscape with song.

Friday 5th

I was seriously overdressed: maybe even two layers too many. Better than the other way round. A robin watched me with frank interest as I took my place. Robins have that way of treating you as an equal: as if we had agreed on something important. Foraging around working humans is an established part of robin routine, as every gardener knows: the robin on the spade-handle is a cliché based on hard fact. When I muck out the stables there is always a robin on the muck-heap waiting to see what delights I have brought; humans value the company of robins, and associate them with condolence in times of bereavement; there are many stories about robins entering a house of mourning.

That thought gave me a small anticipatory shiver as the robin vanished into the ash tree, and I felt a deep shaft of sadness. A while later he began to sing, with contributions from great tit and wood pigeon. The song thrush sang out again, rather more tentative than yesterday. A long-tailed tit perched briefly in the whips growing from the fallen willow, caught in perfect light: tiny, beautiful and just a little preposterous.

> It's generally agreed that non-British robins are much less bold, hiding in deep woodlands rather than flaunting in gardens. That would appear to demonstrate a difference in culture: not the culture of humans but of birds. Or perhaps both.

Sunday 7th

Overnight, spring had been abandoned. I layered up, snood just below my eyes and ears, waterproof woollen hat just above. I walked through the blizzard and took my seat, looking out at the blasted white world with eyes narrow as Clint's before a gunfight.

I was forced back in five minutes. The wind, gusting at 50 mph, was blowing the snow straight into my eyes. I couldn't see with eyes open or closed, and both ways hurt. The northeast wind was dead in front. I could, I suppose, have shifted the angle of my seat, but that would have been cheating. So I went back to my hut.

And returned at once, now wearing, ludicrously, a pair of Ray-Ban Aviators. I have owned them perhaps a quarter of a century and wear them only when travelling at speed in open vehicles, mostly in Zambia. They had never been tested in snow.

I sat there looking at the poor marsh while the lenses,

unequipped with windscreen wipers, slowly grew opaque. It's all very well to look out of the window and gasp at the loveliness, or to do the outside chores at full speed with chilled and aching fingers: but to meet the full force of the weather unresisting was a different matter.

The Cetti's warblers that lived here were killed off by the Beast from the East in 2018 and had failed to return in the last three breeding seasons. Would the Cettis I'd heard in recent weeks survive this lot? If they didn't have good fat reserves and/or a good place of shelter they would die. Those that survived would be tough, smart, well-equipped, lucky, or any combination of the above: that's what evolution means. There is grandeur in this view of life, as Darwin said in the last pages of the *Origin*: and there was grandeur, too, in the marsh spread out before me. A black-headed gull flew along the river, white on white: a vista of the most terrible beauty.

> 🍃 I finally grasped what Darwin was on about when I read the essays of the American palaeontologist Stephen Jay Gould. The series begins with the volume *Ever Since Darwin*.

Monday 8th

The snow was still falling but the wind had shifted to the east, so I could see without Ray-Bans as long as I looked left, or west. A moorhen flew along the dyke in front of me, legs trailing behind like afterthoughts. A buzzard yowled from the right-hand carr.

After a few minutes the wind got up speed and blew snow

into my right ear. There was a brief discussion from the rooks behind me: rooks seem to take everything without drama: keep cawing and carry on. Two great tits flew along the dyke, not a usual place for great tits, but below the level of the bank they were out of the wind. No fools.

A strange thing happens when you look at any form of precipitation through binoculars: you can see what's falling for a good mile. It doesn't become a mush: it remains a series of individual strokes, like looking up close at an impressionist painting. Each falling flake was visible, gentle, remorseless, relentless. No such thing as bad weather, only inadequate fat reserves.

Two crows flew across the marsh, cutting through the vast collection of flakes, black on white, white on black.

> Birds and non-human mammals are no keener on extreme weather than we are. They will adjust their behaviour to find shelter, forage opportunistically, and very often stick together.

Tuesday 9th

The idea was to catch one of the lulls; it had been alternating still periods and sudden gusty sessions of snow. I stepped out in quiet and almost at once I was sitting in a snow-shower. I tried to persuade myself this was the more valid experience.

I sat watching the swirl, for there was nothing else to watch. Then it stopped as if someone had flicked a switch: in the sudden windlessness four ducks got up and flew across and I could hear the strong, uncalm voice of a crow. I could feel each individual flake of snow as it hit my waterproofs: gentle and lethal.

A flock of thirty-odd wood pigeons disappeared over the right-hand carr in a brief, busy crowd: indomitable birds. No one likes them; they don't care. The shooters had given them a prolonged seeing-to last Saturday, but they are Whac-a-Mole birds, mocking every effort to control them. Wish they were all like that.

The snow got heavier. A song thrush perched in the baby oak. The brown speckles might be considered a bit dull in some circles, but in this white world it looked as gaudy as a parrot. A marsh harrier flew over my left shoulder and carried on northeast as if late for an appointment. The snow was building up on my snood. The whole damn place seemed pretty indomitable. I felt pretty indomitable myself.

> The pheasant-shooting season stops at the end of January, but enthusiasts continue with the shooting of pigeons and corvids during February, as 'pest' species. The NGO Wild Justice is contesting the legality of the way the required General Licence is put to practical use.

Wednesday 10th

It wasn't actually snowing. Balmy, then. A buzzard flew overhead: the pale patches stood out so brightly it was as if it had snow in its underwings. The changed colour values of the world had made everything that was not white seem freshly painted: like seeing them for the first time.

Four swans in flight: were they flying towards me? Or away from me? In this foreshortened view all I could see was big chunky bodies and hard-beating wings. At last I could make out four necks and the mystery was solved: they were leaving me heading north.

The dyke before me was frozen. Water like a stone: what of the poor kingfisher? Was there enough for all on the still-flowing river beyond? And then, with an almost smug sense of drama, the sun came out and the light bounced back off the snow all around. At once a chaffinch sang out, though not with complete certainty. I could see my shadow almost blue on the snow, and hear the distant song of goldfinch. I wanted to turn to the camera in the manner of David Attenborough to say, 'And – ever here – there is spring!'

As if that wasn't enough, a jay flew over the marsh, almost sparkling. It was a sight that would have made a King of Saxony bird-of-paradise suffer a moment of self-doubt.

> Jays don't always look colourful, but their flight silhouette is distinctive. Look for sharply serrated wings and movements like a brilliantly operated puppet.

Thursday 11th

That's enough beauty. I'd had my fill of perfect vistas of white and gold. I was ready for normal colours again: normal temperatures, too. A golden hare ran a circle on the white snow, paused, and then did it again: an early hint of March madness.

A buzzard flapped untidily across, reminding me of Leonard Cohen's Suzanne, wearing rags and feathers from Salvation Army counters. So all at once the sun poured down like honey and I could almost imagine that it was a source of heat.

More white on gold: the female barn owl, driven no doubt by a little desperation, was hunting in full daylight. She worked the marsh with concentration, dipping, turning, hurrying. Then she was down: I could see her tearing strips from whatever small thing had mis-stepped that afternoon.

I felt an almost-warm glow for the bird, rewarded for her tenacity, rewarded for her beauty – and then came five whoops of joy, the sound of the male kestrel, homing in on the owl like a jetfighter, a low, shallow, curving dive at top speed. What could the owl do but back down? I hoped it had managed to eat enough to keep going. The two birds vanished in opposite directions.

I sat back, content – but then I caught a flicker, followed it with the bins and it was a kingfisher, perched at the far end of the dyke, perfectly still now and perfectly salmon pink. Then it shifted position and was perfectly blue. An instant later, gone. Not dead, then.

A flock of linnets foraged through the snow, standing out strong against the whiteness, busy, safety in numbers. The barn owl returned: the linnets didn't fly away; instead they melted. Perhaps they had chosen stillness as a tactic. The owl worked the marsh for a while, but didn't catch anything. Perhaps it would get lucky when the sky darkened, and barn owls, not kestrels, are at their best.

> Those personal dramas of nature are not scripted. Sometimes you get hours without very much happening at all; sometimes half a dozen year-making highspots go past in quick succession.

Sunday 14th

There was a slamming cold wind in my right ear and the marsh was as white as ever. My daily outside chores had assumed the proportions of a game-show ordeal. Sitting outside I felt locked up: held tight in the unforgiving grip of this claustrophobic weather.

As if in rebuke a gorgeous female marsh harrier flew across the marsh with the sharp-angled grace that only a harrier can manage: and with this calling to order – how many more will I experience in the course of this journey? – I felt a little shame. She disappeared over the right-hand carr. I looked left – and then realized that the harrier was coming back, riding with the wind in her tail, covering the quarter-mile between the two carrs in ten effortless seconds. A lone lapwing flew over.

And then one of those familiar and yet oddly shocking dramas: two marsh harriers came together in the air – not sure where either of them came from, whether both were female or whether one was a young male – and for a moment they were talon-to-talon above the snowy landscape. At once they parted and flew off in opposite directions. Here was a prevision of the great spring skydance of the marsh harrier: one of the greatest sights that this country can offer.

A snipe rose sharply in front of me, for no good reason that I could make out. The cold weather usually brings in a few snipe: this one turned sideways in flight to give me a good look at the beak, one apparently borrowed from his big brother. Behind him four more lapwings.

> Whenever you see any living creature doing something inexplicable, it's almost certainly something to do with sex. The great game of outwitting rivals and attracting the finest mate – or mates – can be viewed in a million tiny nuances and also in the greatest dramas in nature.

Monday 15th

Green! There were still a few patches of snow, but green was now the dominant colour. For this relief much thanks. The soundscape had changed too, but subtly: no instant detonation of song. It was an afternoon of slow eagerness.

A jay flew across and perched in the clump of sallows, screeching loudly. Why? It moved on into the heronry. Two herring gulls flew along the river, yelping excitedly to each other: you don't need a beautiful song to express the power of spring.

A dunnock started to sing close by, a robin did so above my head. A stock dove started up. Then another jay appeared, also calling hard, like the scream of machinery in torment. Explanation: the jays were working as a pair, keeping in touch by the highly effective means of their voices.

It was good to think that everything was either in pairs or trying to be.

A little egret flew along the river; I remembered old television adverts for soap powder: your whites make my whites look grey. And then a not-quite treat: two geese flying overhead, calling indistinctly. No orange on their beaks, so not greylags. Bean geese? Pinkfeet, escaped from the flock? Time for a rueful smile: a better birder would have known, even in this fading light. But a little owl called, a chaffinch sang – and that was something, was it not?

> A shortfall in the skills department is no bar to enjoying nature ... but greater knowledge and experience brings its own rewards, often in the form of birds.

Tuesday 16th

Spring was creeping on, defying the odds. The dyke before me was still iced over and there were patches of snow on the banks, but the air hummed with intermittent song. The extreme weather was over, at least for now, but it was still a little daunting.

Conscientiously undaunted, I sat down, finding three shelducks on the ground half a mile off. Shelducks gather in small groups in the corners of fields and stand around for no very clear reason, their bright colours, white and chestnut with red beaks, making a sharp contrast with their surroundings. They often seem to be operating as pairs within a flock.

A chaffinch sounded with sudden boldness in the willow to my left, going through his unchanging song a dozen times. No: there were six shelducks: three more had shuffled into view. Another flew overhead, turned economically and parachuted down to join them.

A dunnock decided that now was the time and let rip with a series of cheerful phrases; a blue tit and a robin added contributions of their own. Then a wood pigeon.

It's coming.

> You can broadly divide songbirds into two types: repertoire singers, who are inventive and various, and stereotypical singers, who sing the same thing again and again. Chaffinches, with a phrase that's often compared imaginatively to a fast bowler's run-up and delivery, are emphatically of the latter kind.

Wednesday 17th

A pair of mallards rose from the dyke as I approached my seat. Apologizing, I sat, the wind at my back and the sound of a buzzard overhead. It was cloudy and cool.

A little owl called and startled me out of a train of unproductive thoughts ... how does this happen? Sometimes not thinking feels like the greatest luxuries of our time. *Non cogito ergo felix sum*: would Mr Neath, my fifth-form Latin teacher, approve that sentence? A great tit sang briefly; a robin kept going; the buzzard had never stopped. But you have to be able to do both, don't you? You must think, but you must also find ways to escape thought. Nature offers opportunities for either.

And then one of those perfect moments, and yet again it involved a marsh harrier, and I was once more the Zen master with my brush in my hand and a virgin scroll before me. Just three strokes, a broad and crooked one for the willow stump, and, change the brush, two more of supreme delicacy for the harrier's dihedral: as if my own clever fingers, my own adroit wrist had created the picture before me – and now it was gone, all the richer for its lightning-brief existence.

After a sprinkle of rain the sun threw a real shadow, one that lasted for at least five seconds. I even fancied I heard a fragment of a call from an oystercatcher: birds that come here from the cold coast for a spot of breeding.

> It lasts a second or two of objective time: it stays with you for ever. True of many things, but especially of the pursuit of wildlife. Here's Blake again: the hours of folly are measured by the clock, but of wisdom no clock can measure.

Thursday 18th

Don't move. Stay invisible. Every time the muntjac raised his head, I froze. When he lowered it I took another pace or two towards my seat. I've played this game in the playground of Sunnyhill Primary School, also in Africa, where it's even more exciting. When I got within 30 yards the muntjac retreated: a dapper beast after the Chinese water deer.

A barn owl flew over the marsh: the place still had plenty

of food then. Perhaps last year had filled the place with short-tailed field voles, favourite prey of barn owls and kestrels. Already they would be getting down to some enthusiastic breeding.

Two hares were feeding. Earlier in the day I had seen three of them, chasing each other in circles as they anticipated the madness of March. Now the situation had been resolved in some way, and food is always good. Like all grass-eaters, hares need a good through-put.

And what a sky. All smoky blues: it might be beyond even yesterday's Zen brush-master to capture it, mixing the fierceness of winter with the fierceness of spring.

The sun came out just to show that it could, and the trunks of the alders looked greener than I had ever noticed before. A sudden bark from distant deer.

> Short-tailed field voles are the commonest wild mammals in Britain: there are about 60 million of them. A female can produce six litters in a year. We humans don't see them much because they tunnel through tangled grass. Barn owls car hear them, though; and get a cross-bearing on them from their asymmetrically placed ears.

Friday 19th

It was cold and windy, with plenty of high cloud. Two buzzards were soaring, at ease among the buffets. What's distance to a buzzard? With one flick of the wings either of them could travel half a mile with no more muscular effort than it takes to reach for a cup of tea.

Behind the consistent roar of the wind a great tit sang its two optimistic syllables again and again. A thrilling bird, a lanner falcon at the very least, came fizzing across the marsh: in this wind even wood pigeons are exciting.

There were buds on the ash: how long had they been there? Black on grey, a long way from bursting, but getting there. And then the marsh was full of mallards: more than forty of them flying across diagonally in half a dozen bunches, calling hard.

A marsh harrier followed them a little later, adjusting and readjusting. Tough weather: spring advances and retreats, advances and retreats: but like an incoming tide the advances are always a little bit farther, the retreats a little bit less.

> Pigeons can be mistaken for many far more glamorous species, though seldom for long. Feral pigeons, being more various, have the greatest scope for confusion, but a silhouetted wood pigeon – when you can't see those white patches on neck and wings – will often bring a moment of excitement.

Monday 22nd

The rain was light, the weather still warmish after a sudden outbreak of spring the previous day. A female mallard gave a loud call of approval. A robin and a great tit sang, and then a wood pigeon. Another bird called from the ash: and it wasn't a great tit, not quite. That should have meant it was a great tit, of course, under Bill Oddie's Law, but I still didn't think so.

There is a deep pleasure in mysteries and a deep pleasure in the rare occasion you solve them: and both pleasures seem to me deeper when they concern the wild world. A song thrush struck up, singing with more confidence than last week, stringing more repeating phrases together. This one tried a rather good phrase, paused to savour the effect, and then tried another.

Ha! It was the call of an oystercatcher, sweetly mimicked: song thrushes love to borrow musical ideas when they occur in their arc, the oystercatcher's valley-filling piping is just the job, and it's in the repertoire of the local song thrushes. The oystercatchers had yet to arrive but their song was here before them.

And then a sneeze. A rather ladylike and refined sort of sneeze: and that's a giveaway. Another mystery was solved. The bird in the ash was a marsh tit – and there it was, clinging to the lower twigs in its sleek black cap. It was like doing a jigsaw puzzle, except that it wasn't bits of wood fitting together but an ecosystem. For a fraction of a second I felt like Charles Darwin.

> Female mallards make the classic *quack-quack* we attribute wrongly to all of duck-kind. Male mallards go *quirk-quirk*. The marsh tit's sneezing call is described as *pit-chew* in *Britain's Birds*.

Tuesday 23rd

The firm wind at my back told me that next-door's muck-spreading had been an unqualified success. Two robins and a blue tit sang out to express the subtle warmth of the day; the soundscape has changed radically over the past few days.

The sun came out and two shelducks flew up as if to catch the light and show off their feathers at their very brightest. A dozen black-headed gulls flew over in a ragged, undisciplined bunch, neither travelling purposefully nor foraging in the air. They seemed slightly at a loss as to the best thing to do. A common gull broke the pattern, flying low, giving me a small moment of pleasure as I picked it out.

Two or three chaffinches exchanged contact calls: the sharp sound that gives them the second half of their names. And then two buzzards flew across. The lower one twisted a wing and in an instant gained 30 feet of height, revealing the astonishing pallor of its underwing: one of those sudden shafts of enlightenment that nature hands out so profligately to us all, insiders or not.

> Common gulls are not the commonest gulls in this country and are much overlooked. Look for large white patches within the blackness of the wingtips.

Wednesday 24th

I was glad to sit down, woozy from my Covid jab, accepting rather than resenting the inconvenience. Safety, sheltering people like Eddie and my father, it's a shared responsibility, is it not? I had gone to the vaccination centre with Cindy and Eddie: we had made a brief detour to Horsey Mere, looked at a flock of greylag geese – and it seemed like an expedition to Tierra del Fuego.

Above my seat two robins and a great tit were singing; a fairly distant song thrush threw out an occasional phrase. The day was cool with high cloud. There were two buzzards in front of me; one made a vertical descent with horizontal wings, a very steep glide rather than a parachute-drop. Further off, a marsh harrier was soaring, gaining height rapidly and rising beyond the reach of the naked eye. This is not everyday behaviour for harriers. I focused through the bins as a descent began, realizing with a start that there were two birds up there. They circled each other a couple of times and then separated.

Too far to be sure, I suspect that this was a female with an adult male. If so this was a significant change to the dynamic of the marsh. The two of them had, I believed, just performed a restrained and dignified *pas de deux*: precursor to the full skydance that is an essential part of their courting. This was more of a sizing-up: eyes meeting across a crowded sky. Were they strangers? Or did they know each other of old? Had they made marsh harriers together last year? Or was this the start of something new?

I heard the sound of herring gulls and saw three of them together, as high as the harriers had been. No – two of the silhouettes were the shallow Ms of gulls, but the third was straight, the wing long and whangy. Red kite, then. Will we have breeding kite this year? These birds of prey and their flying styles: kite for the all-day glide, buzzard for the soar, harrier for control at low speed.

> You can learn to recognize birds from the way they fly. It becomes an unconscious thing, like recognizing a footballer from his movements on the pitch. Familiarity gives you something that can seem to novices like extra-sensory perception. But it's a knack easily enough acquired.

Thursday 25th

Every Saturday in February they shoot up the heronry, killing pigeons. The herons aren't in it when they begin, but there's always an overlap at the end. Every year I fear that the herons will abandon the place because of this disturbance. Every year – so far – they haven't. But I have grown to hate the period of uncertainty between the shooting and the nesting.

And as I sat – the day cool and cloudy, two robins and a great tit singing, and one drummer drumming, this being a great spotted woodpecker – a heron skirted the left-hand side of the heronry, apparently making a recce. One more Saturday left, though.

A slim white neck snaked above the distant dyke by the line of sallows. Swan? Surely not: and it revealed itself as a little egret. One day, perhaps, a pair of egrets will move into the heronry.

You can spend a lifetime watching wildlife and still see things you've never seen and can't explain. A herring gull flew across and dipped sharply down – and below it was a hard-cantering deer. The gull was chasing the deer, flying at top speed as the deer ran on, until it was lost to sight in the reeds. The gull rose up and went back the way it had come, in a more leisurely fashion. It looked like a piece of pure horseplay.

A movement high above the marsh drew my eye: a bird dropped down onto a crow that was flying beneath, forcing it to retreat. It was a kite, in no mood for mobbing behaviour. It was at once clear why: there was a second kite. Very slowly and very beautifully they circled each other, as if playing first one to flap is a sissy. Holding the height, maintaining the same distance apart, they made circle after circle, drifting away to the northeast. In the five minutes I had them in sight I counted a single flap. Hope that doesn't count against you.

> What are they doing? Why are they here? These are questions that enrich your experience of wildlife, and take you beyond the issue of identification.

Friday 26th

As I sat out on a cool, cloudy day, weary to the bone from my jab, the sun broke free and its golden light rushed towards me. A harrier was hanging in the air above the distant oaks, a dunnock was singing, and close by, on the top of a tangle of brambles, a wren sat with all the perkiness that wrens can manage.

A kestrel flew into the lone alder, now established as a favourite hunting perch. After just a couple of minutes he dropped: a very rapid and steep glide, wings pushed back, though not quite the full anchor-shape of the stooping peregrine. That was the last I saw of him that day: menacing silhouette followed by the disappearance that indicates death on the marsh.

And then there were two blue tits among the catkins of the coppiced hazel the far side of the fence: such prettiness, such daintiness, such charming birds among the sunlit golden catkins. They were, of course, on the same business as the kestrel: working their way through the catkins killing as they went, taking small scraps of invertebrate life as they built themselves up for the great event of the year. A robin sang in the ash above my head: as I listened my weariness seemed something to savour.

> It helps to have more than one way of looking for food. Kestrels are most visible when they hunt from a hover. Hunting from a perch is more economical, but less versatile.

Sunday 28th

Let's call them Hallelujah Moments: milestones you pass during the advance of spring. The sound of oystercatchers – real ones, not mimicking thrushes – filled the valley as the piping hooligans returned. They were not acting cool about this achievement: everything in an oystercatcher's life is the most exciting thing that has ever happened.

A hare cantered slowly down the mown path on the marsh before me, not exactly worried, but easier in its mind when it had put an extra 20 yards between us. The ears, black-tipped, looked like feathers, but everything about a hare is slightly improbable.

Then came the two most longed-for notes in the world. It was a chiffchaff singing his own two-syllabled song. Chiffchaffs are the first migrants to arrive in this country, though I suspected that this one had over-wintered here, as many do these days. He had survived the frozen week and was now able to lift his voice ahead of all those who had preferred to winter in southern Europe.

At first the song was distant and hard to pick out with certainty, like a flavour in a good malt whisky that only the person who wrote the label could taste ('hints of lemon, toffee and chocolate in the finish'). But gradually it made itself good and clear and strong: a ringing solo like the heart of the malted barley itself. Raise a glass to spring.

Away in the boundary dyke I saw the egret again – and then there were two of them working alongside each other, making a big thing of being together. If you're looking for a nice heronry I know just the place, for now the time of shooting is past and the voice of the chiffchaff is heard in the land.

> Chiffchaffs and great tits both sing twin-syllabled songs, but they're very different, and not just in timbre. The great tit stresses the first syllable, hence *teacher-teacher*. The chiffchaff gives an equal stress to each syllable: *chiff, chaff, chiff, chaff* ...

MARCH

Tuesday 2nd

I took my seat with the air of an old man lowering himself onto a deckchair, still full of post-jab weariness. And did a double-take: there was *green* stuff on the bank of the dyke. Actually growing. The nettles were coming back, not the most welcome harbinger of spring but an honest indicator nonetheless. Nettles were once prized as the first edible greens of the year; the stings vanish in the cooking. Apparently.

A dunnock, perhaps the best of the year so far, struck up along with robin, wren and great tit. A wood pigeon sailed across the open space with a single clap of self-applause. He would set up a breeding territory here unless some other bird staged a coup.

The buds were thicker on the ash, the sharp outline of the branches was blurring. In the winter the branches form faces when you stare at them: that morning I had looked at the ash from my bed for the familiar features of D. H. Lawrence and Charles Darwin, and they were beginning to vanish.

I scanned the new vegetation through the bins to see if anything other nettles was growing and found the Velcro-sticky green stars of cleavers. The two egrets flew in formation along the more distant boundary dyke and went down together behind the sallows.

> Cleavers are easy to recognize from their stickiness alone; they also have distinctive square stems. They can grow ten feet in a single growing season.

Wednesday 3rd

Spring was in full retreat. On this cold, overcast day I could hear no song thrush, no chiffchaff. Even the great tits had gone quiet. There is a special desolation about such days: it's as if spring had missed its trajectory, as if the seasons had given up, leaving us all betrayed, hopes forever dashed. The only real sound of spring was the gas-banger from the neighbouring farm: a bird-scaring device that sounds like a shotgun fired down a well, keeping birds away from young crops.

Still weary, I leant back and closed my eyes. I could hear buzzards over the heronry and the derisive yells of black-headed gulls. And then all at once a robin struck up and sang and sang and sang and there was nothing I could do but listen, sitting, listening as the song unfolded.

To the human ear the song is wistful, gentle, tinged with a soothing sadness, but for the singer it was the song of life, a trumpet voluntary, a military bugle, the Rite of Spring and the Hallelujah Chorus. It was the Song of Songs.

The oystercatchers called some distance off. Two jackdaws were deep in discussion in the nearby meadow; their *jack-jack* calls were softer and more sibilant than their usual harsh din. This was perhaps an operatic aria, a duet of love, *la ci darem le ala* ...

> Bird call – call as opposed to song – is not a mechanical on–off response to situations. It's a strongly nuanced thing, varying in intensity and urgency and no doubt meaning.

Thursday 4th

It was colder, if anything, light rain on a wind that was gusting hard into my face. Rooks cawed; a dunnock sang briefly to my left. A marsh harrier followed the line of the river. A buzzard rose into the wind to my right and seemed to be struggling. Mind you, so was I: I found the snood in my pocket and arranged it below eyes and ears. Better, much. The marsh harrier returned, gliding faster than the wind and the wind was going some.

But I sat on without feeling even slightly brave. I sat on because sitting on soothed my soul; I sat on because sitting on was better than not sitting on; I sat because you never know what will happen next.

And that's when a bird came dashing over the marsh from the far side of the ash and performed a high-speed low-level run. Not a sparrowhawk: this bird was the steel-grey colour of Dirty Harry's gun. As the bird performed a sharp climb on the far side of the boundary dyke the wings seemed to lengthen and sharpen to an impossible degree and I knew I was looking at a peregrine.

I was still marvelling at this piece of luck when twenty birds flew over in a tight but informal formation and circled the marsh twice. As they came down to land they dropped forty long legs and revealed twenty long decurved beaks. Here were curlews, on their way from the coast to the high tops further north, dropping in here to rest and refuel.

> Experience has taught me never to stop looking. Nature gives us more than we deserve – if only we let it.

Friday 5th

The sun hadn't made the place much warmer, but it was welcome for all that, and certainly it moved a dunnock to song.

Almost. It lasted about three notes. He tried again and then again, but those three or four notes were the best he could do: a half-arsed song for a half-arsed day.

Besides, the wind was getting up and the clouds were moving fast, the darkest now right over the middle of the marsh. A buzzard was flapping hard, eventually taking station under the black cloud – where I then noticed a second buzzard.

That's when they both got into the swing of things, using the wind to aid their ascent. Once they had enough height they abandoned powered flight and soared, tails fanned about as wide as they would go, wings spread and stretched, all to create the maximal aerofoil surface. Each bird had become a single wing and soared. Up and up they climbed, hard to find now even in the binoculars, black birds in a black sky, perfect shapes in the sky-reaching lenses.

Higher, higher: I sat there in my chair in a perfect moment of soar-envy: bum in the place it had been resting for damn near six months and soul up there with the soaring birds where it has been for most of my life. As I stared on into the blackness of clouds I could hear the piping of the oystercatchers.

> 🌿 Dunnock is an important song to get clear in your mind: a merry jumble of notes that sound a little flat, but actually aren't. They will sing occasionally in late autumn and bright days in winter, but early spring is their best time. Check the song out (search 'RSPB dunnock', for example) and in season you will hear it all the time.

Sunday 7th

The moorhen went skedaddling from the dyke as I arrived: how can you tell a moorhen you mean no harm? It was warm and sunny and I was two layers slimmer than last time.

The long strip of earth turned over by the digger all those weeks back was greening up a treat, mostly nettles but who's quibbling? A robin sang; a wren took a perch on the summit of a log on the far side of the dyke, splendidly proprietary.

I heard a chiffchaff for the first time in a week, and from the oak to my right a great spotted woodpecker called. Two oystercatchers flew over the marsh: astonishingly, without making a sound. I fantasied a couple not talking after the most hideous row, then told myself sternly to leave off anthropomorphizing.

In the bright cheery light I could see that the lone alder was turning red, as if it was autumn: but these weren't leaves; they were male catkins. I should have noticed them before but never mind: I noticed them now. Four cormorants flew over in tight formation, alternating flaps and glides.

> Alders tend to grow in clumps in wet places. Flooded sites are no problem to them; they just grow on and resist rotting. They also have an unusual skill: their roots attract bacteria that extract nitrogen from the air – which means that unusually they flourish on soils lacking nitrates. The larger male catkins start to appear in January and carry on doing so until April; female catkins, on the same tree, are very hard to see.

Monday 8th

Another of spring's little retreats: my abandoned layers were resumed as I sat in a light rain that was getting less light every minute. A muntjac, grazing in the middle of the marsh, was soon off despite my sneaking approach to my seat. At least the kestrel perched in the oak sapling paid me no mind.

And then all in a single whooshing moment the air was filled with the glorious life-affirming sound of quarrelling: growling, clattering, yammering. My heart lifted at the din: the herons were back in the heronry.

A heronry is not a utopian commune. The birds need companionship, shared information and safety in numbers, but they also resent it. They build and restore huge straggling nests in the treetops of the alder carr, each one carefully out of reach of the snake-like necks and dagger-bills of their neighbours.

Perhaps quarrelling with neighbours is a basic human right, or a basic ardean right, as herons would see it. They were no doubt pinching each other's best twigs, establishing the right sort of social distancing and generally getting ready for the task ahead: which was making more herons.

The rain got harder; the chill penetrated my defiant layers. We think of spring as a time of unambiguous joy, but it's much more complicated than that. Spring is a life-and-death struggle: the toughest time of year as well as the best.

> There are heronries in cities, easy to visit and giving rewarding views. Londoners can find them in Regent's Park and Battersea Park; there are also heronries in Queens Park, Manchester and Cleeve near Bristol.

Tuesday 9th

A robin sang to welcome the newly arrived sun as I sat. It wasn't a great burst of sun, but I welcomed it too: it seemed to imply that the world was doing its best in trying circumstances.

Away on the far side of the river a tractor was ploughing the field in long strips; even at this distance I could hear the grumble of the engine. A well-organized queue of gulls stretched behind it, as if waiting to buy their tickets. Up they all went as the tractor made a headland turn, down they all came again.

The background sounds of tractor, rooks and jackdaws were broken by a chiffchaff, singing for a sustained period. A moorhen called from the dyke in front of me, very close but out of sight. The latest quarrel in the heronry was clearly audible: I leaned forward to look for a snaky neck emerging from the canopy: at this the moorhen gave a squawk of protest and flew off down the dyke, though moorhens are seldom eager to fly.

When I sit here there is no barrier: instead, I am hidden by my stillness. I had broken the code and radioed apologies to the moorhen. Four greylag geese flew over, their honks wavering into a kind of diphthong.

> Stillness really does work. After I had sat quite still for an hour in a Zambian ebony glade, a male bushbuck – a lovely spotted antelope – walked within five paces of me without realizing I was there. Stillness is one of those skills that requires no skill at all.

Wednesday 10th

The wind was at least behind me, firing bullets of rain between my shoulders and using my hat as a target. Before me the trees danced. The willows did it best, waving their arms in the air like hippies at Woodstock.

There was a little courageous singing above the low roar of the wind: a great tit was at least having a go and a robin tried out a couple of phrases. Huge circles appeared and reappeared on the waters of the dyke: the ash tree was collecting rain and releasing it in the form of cannonballs.

Then I picked out two harriers over the reeds in front of the river: that deeply familiar shallow vee, their gliding easy and confident even in forbidding circumstances. They too appeared and disappeared, hard to pick out, pale grey shapes against a pale grey sky. The other day I had flushed my pen with water to ease the flow of ink: when I used it again the marks I made were first invisible and then in a grey so pale they almost weren't there.

And so the Zen master of calligraphy sat there beneath his ash tree planning a long, perfectly off-white scroll: all you could see on it – and you'd have to look closely – would be two vee-shaped marks in perfect off-grey.

> 🍃 A very little basic tree knowledge will deepen anyone's appreciation of landscape and the nature of landscape. Willows love watery places, especially the banks of rivers and streams; their long, think whippy twigs move with grace at slightest provocation from the wind.

Thursday 11th

A rainy morning had given way to an afternoon of intermittent sun, but the wind was still hard at it, coming in sudden 40-mph gusts which filled the valley with a tube-train roar. It seemed as if the whole world was wind, but I could still hear little moments of long-tailed tit, great tit and robin.

Timbre. Whenever you're not quite sure what you mean, borrow a foreign word and use it with confidence – or better still with panache or élan. Because I heard a robin that wasn't quite a robin. Or was it?

There's a difference in timbre when you play the same tune at the same volume on, say, a church organ and an electric guitar – or more subtly a clarinet and an oboe: same melody, different sound. I sat and waited, hoping for a repetition, knowing that sometimes you're mistaken and sometimes – well, you're not.

And for once I wasn't. The bird sang again and it was the bullfinch: gentle musical muttering from deep cover. These splendidly gaudy birds seem embarrassed by their own beauty, shy of exposing their deep-flush pink. They hide, they sing from cover and when they do so they sing softly. They must be the most overlooked bird in British gardens, mine included.

A bank of black clouds raced over the marsh, hiding the sun. A bird appeared from the left with such grace that at first I couldn't identify it: then it changed its shape in the air and became slightly awkward and an obvious heron, dropping down from the heronry to refuel. Beyond the sallows two oystercatchers were busy in the short grass.

> 🍃 Once you have a few birdsongs logged in your mind you open yourself up a new pleasure: the thrill of finding a song that you don't recognize. You try and hold it in your memory, perhaps invent a mnemonic, and then try a few possibles, from computer or phone, by means of educated guesses. And every now and then, you solve the mystery and rejoice.

Friday 12th

The sky had divided itself in half: on the left a cheerful blue, on the right, uncompromising black. Deep in the blackness I could see a black-headed gull, not like the Ancient Mariner's albatross but the one that guided the *Dawn Treader* and its crew of Narnians away from the dark horrors of the Land Where Dreams Come True – the one that whispers 'Courage, dear heart'.

The wind was more vigorous even than yesterday. A jackdaw was adopting the special flight they go in for in such conditions: when they appear to be tossed out of control by the ferocity of the wind, blown about like a bin-bag – but all the while they are using the wind with purpose, as a white-water kayaker uses the frothing water.

Then a double change: a little rain came and with it the sun. A goldfinch sang softly, a golden song appropriate for this moment of golden weather. The gabbling sound of shelduck came towards me from beyond the boundary dyke and two of them rose and flew energetically around, catching the light, shining white, newly washed. Round they went, again and then again, together, always together.

> It's always nice to see a bird like a shelduck that is so uncompromisingly straightforward to identify: a bird that gives reassurance in a world of doubt.

Sunday 14th

You never know what will turn up. I know that sounds like insufferably bracing optimism, but it's the literal truth. You know you'll probably see only the sort of stuff that's usually about, and what the hell's wrong with that? You know you might have a poor day with very little and console yourself with the thought that negative data are still data. And then of course you might just find something special.

It was a bright day, cloud and sun taking turns and a sharp wind. A whisper of bullfinch, then of robin. Beyond the boundary dyke two – no, three – swans, widely separated. Rising just above the dyke, the serpentine neck of a heron – hang on…

A slight change in the light and a shift of position from the owner of the neck and I could see nothing but white. Then the bird clambered out of the dyke to make it unassailably clear that I was looking at a great white egret, only the second time I had seen one here.

Its considerable size – as tall as a grey heron, but slender and more graceful – was made clear by the strategically placed swans. The beak was yellow and not black. The bird stood there for a while, looking like a beautiful woman modelling a silken gown in the wilderness for a gimmicky fashion shoot, before stepping delicately back into the dyke.

The curlew flock, thirty-strong, rose from the same field, where they had been invisible among the swans, performed a circuit and dropped back into invisibility. Two harriers showed up, very briefly and indistinctly. One flew left and one dropped down more or less directly in front. Significant spot?

> Ah yes, rare birds. Seeking rarities is the breath of life to some birders, and good luck to them. For others, the pleasures are less dramatic: a closeness to nature that comes from intimacy with the ambient birds of a place. I make no moral or aesthetic judgement: any way you enjoy your birds without killing them is fine by me. The top seekers of rarities are mostly a hundred times better than me at species identification.

Monday 15th

How high? How can I estimate the height of a flying bird? It was certainly a long way up. The day was sunny and the wind still vigorous: the buzzard, travelling fast, turned into the wind to make an instant halt, like a show-off skater: I remembered them from boyhood visits to Streatham ice rink pelting at the gate at 100 mph before stopping dead in a knee-high blizzard of ice.

The sky was full of Magritte cumulus clouds; robin, chiffchaff and dunnock in song. Two herons flew down from the heronry in tight formation and then, as if by prior arrangement, separated to different ends of the dyke.

I looked up in response to the calls of buzzards: four of them almost directly overhead, climbing in a thermal. I had once ridden in a glider in just such a thermal: not as peaceful as it looked from the ground, noisy, buffeting, bouncy, alarming and thrilling.

Up they went. There was a fifth bird among them, surely a kite – yes. So how high? The lowest buzzard looked much the same size as a buzzard does when perched on the lone alder: so – brilliant deduction – the flying bird was as far away from me as the tree. So later I paced the distance between tree and chair, estimating for the two unleapable dykes. It was 170 paces, say 500 feet. The buzzards had carried on climbing far beyond 500 feet until they were lost to sight.

> Thermals create opportunities for big gliding birds and for those who watch them. At some important migration spots thermals will be filled with hundreds of circling birds of prey and/or storks: they use the height they have gained for free to travel onwards.

Tuesday 16th

It had been an intense three hours of work to a tight deadline: once done, in the quiet satisfaction of craftsmanship, I turned with enthusiasm to my deserved reward: a nice long sit on my favoured seat. It was only as I was layering up for this treat when I realized that by any sane estimation the day was foul. It was cold and the rain was blowing horizontally in classic Norfolk fashion.

It wasn't just that I was undaunted: I was genuinely delighted to sit down in this stuff. A subtle change had been working on me over the past weeks and now I was aware of it. I sat almost gleefully. After a while – quite a long while – I heard the voice of a robin. An actual bird. Song duration, approximately 1.5 seconds. And was that a moorhen? Hard to tell in this cacophony of wind.

But then unmistakably – perhaps it was exactly the right timbre for cutting through the wind – the lovely note of curlew: so they were still around, waiting for the moment to set off for the high country. A moment later a harrier cruised softly over the reeds, so perfectly controlled you'd have thought the air she flew through was still and perfectly dry.

> Curlews seem often to be associated with wild weather: their calls bring out the romantic in all of us. The name is onomatopoeia, but they also make a longer, heart-rending, bubbling cry.

Wednesday 17th

Lord, I was weary of the wind. I felt confined by its persistence: it was making this vast open vista with its towering skies seem like an enclosed space. There was no escape: inside you could still hear the roar and see the restless trees: outside it whipped my face. A dunnock sang with wild optimism, geese honked from the Flood. The sun came out and threw down a token of warmth before the wind swallowed it up again.

And there was a kite crossing the marsh with the wind

in its face. One of the skills of paddling a canoe is to read the gusts before they hit, so you can make a pre-emptive manoeuvre: the kite failed to do that and was blown backwards for a few yards. Unworried, it dropped into a shallow dive to regain airspeed and so move forward again relative to the ground. Once that was established it used the wind to climb: all this, of course, without a single flap.

A mad screaming made me turn my head: one black-headed gull was in full pursuit of another, the wind in their tails. The wind was as exciting to them as each other's company: they were like young lovers idiot-dancing in the sitting room. Two shelducks crossed the marsh with the same air of being totally caught up in each other. Five minutes later they were back, much flapping as they turned into the wind – they're no kites – and then, with the wind behind them again, they went barrelling off. In their wake a little egret toiled sedately.

> Windy conditions will keep a lot of birds under cover, so it pays to keep an eye on sheltered spots. The same conditions prompt other species to put on a show.

Thursday 18th

It was slightly less windy, but to make up for this, the rain was back, coming in sharp, painful gusts. But before I had a chance to complain a Cetti's warbler gave a glorious unmistakable Cetti shout. It was the fourth time I had heard him in three days, and the first from my seat.

So it was a good day already as a kestrel slashed its way

through the wind and went into an energetic hover: the alder altimeter put the height at 100 feet. She held station for a good three minutes, altering her body shape in the wind but maintaining precisely the same position relative to the ground.

Then, dissatisfied – had the prey she had lined up shifted, gone into hiding, become aware of the aerial menace? – she side-slipped away and resumed the hover 50 yards on. After a moment she shifted again – so much energy was being used up – and then went into a sharp stoop to the ground. A pause: and then she was flying into the lone alder, nothing in her talons but air.

After a while she moved to the oak sapling, but when a buzzard appeared she took off again, uncomfortable in this exposed position. She perched again in the sallows, looking tousled and rather bothered in the wind and rain. Kestrels almost always look suave: there was that rare thing, an uncool kes. As I looked at her I realized that the sallow was almost in leaf.

> Cetti's warblers are classic climate change birds: once sought-after exoticisms, now ambient birds of wet places in a good deal of England. Little and great white egrets have also increased with the shifts in climate.

Friday 19th

It was a beautiful day. I took my seat with joy and at once I heard the Cetti again. Then I heard the roar of power tools:

Cindy was gathering hazel wands from the marsh, working on a new project for the Raveningham Sculpture Trail. So I took a pause.

I was back later than usual, then, five o'clock, after Cindy had finished. In the lush orange light the place was full of song. A series of barks, growls, yaps and grumbles told me that the heronry was doing well: an ugly sound with a beautiful meaning, one that makes the sound itself beautiful in the knowing ear.

A tiny movement caught my eye and for a tiny moment I saw a lynx. Say a quarter of a second. It was the tips of the ear that did it: black and visible above the grasses and sedges of the marsh, the rest of the animal hidden. Hare of course – no, *hares*, for I could count four ears and am fully capable of dividing by two. They made the ungainly hare lollop to a more open spot – hares are only graceful at speed – and they fed together in the sun, their lush orange pelage almost glowing.

These black ear-tops are surely signalling devices. When one hare can see the ears of another moving slowly, it's clear that there's nothing to fear. When they stop, erected, it means the owner is fully alert, suspecting trouble. When hares are together they keep an eye on each other's ears – and if they see a lynx they run. Gracefully.

> Lynxes are part of the ancient British fauna, but went extinct here about 1,300 years ago. There are groups like Rewilding Britain campaigning to bring them back. My guess is that if they returned few people would notice: lynxes are very good at keeping out of sight.

Monday 23rd

The equinox had taken place at 0937 the previous Saturday, so this was my first official sit in the official spring: the first time in the course of this project when the daylight hours outnumbered the hours of darkness.

It had been a classically gorgeous morning, and though by the time I sat out it was cool and cloudy you could still tell it was spring with your eyes shut. No cacophony: instead a mutter and mumble of song from all around, carrying on as if it had never stopped. I broke the murmuring into its constituent parts: blue tit, great tit, robin, chaffinch, rook, magpie and wren. Then, very loud and very close, a Cetti shouted his best shout.

Two harriers were manoeuvring over the reeds: no chance of sexing them, for they were just silhouettes, and too far off to give an idea of relative size (females are bigger and burlier). They rose up to about 300 feet. At this point I lost one, but the second caught the light and with her pale head she was unequivocally female, finding some lift and rising in tight circles, tail fanned. She stayed with the thermal until I lost her at about 1,000 feet – two alders.

A movement close by and I shifted focus: a dunnock among the hazel catkins, looking like the most beautiful bird that had ever hatched, a symphony of brown and ginger caught in the diffuse sunlight of spring.

> 🌿 I have sometimes been told that knowing and naming the individual species that sing out on a spring day is soulless and pedantic. It is no more soulless than knowing the instruments of the orchestra when you listen to Beethoven's Ninth.

Tuesday 24th

There was a fuzz of almost-leaves on the more distant sallows. The day was cool, bright and cloudy. A marsh harrier went past and I had one of those sudden unreasoning moments of glory: not gloriously unfamiliar like the great white egret, but even more gloriously familiar: oh brave old world that has such creatures in it. She moved on, and then as she came back she seemed to light up, her pale head almost white; for a moment I was looking at a bald eagle.

A chiffchaff sang out: there is tendency to slight impatience at this time of year as you wait for the later migrants to arrive: as if spring was marking time. But it's doing nothing of the kind: the residents are at it hammer and tongs, and so are the chiffchaffs, at least some of them short-haul migrants. A blackbird sang out and was immediately challenged by another.

Then the sky fell in. And surely the temple veil was rent in twain as well. The world was filled with thunder: a noise that went from horizon to horizon. It could only be some monstrous military aircraft – Norfolk is full of such things – but I couldn't see it, even though it was doing a totally convincing imitation of the wrath of God. It went on for so long that I began to doubt my sanity and the solidity of the world I was looking at.

At last it was gone. A buzzard flew by at 300 feet. Silently.

> Many people know blackbird song without being aware of their knowledge. It's the laidback whistling that comes down from the chimneypots in spring. If you can get that song into your head in March you will hear it again and again over the next two months.

Wednesday 25th

The wren was so loud and so close that it startled me. The day was cool, the sky lightly covered by high clouds. And I was worrying. Just a little, but worry is part of the deal when you get caught up with nature. When will the migrants arrive? Will they come in proper numbers? Is the Cetti still here? Are things OK in the heronry? I live with wonders: if the price for doing so is worry, then I pay it without grudging.

A heron stepped out from the heronry to perform the classic arch-winged glide from the top of the alders down to the dyke 300 yards away. Good. A chiffchaff sang out, got into its stride and gave it everything. Birdsong is an honest indicator of a bird's health and mate-worthiness: you can only sing loud and long if you're in top condition. Song proves you are up to raising a brood: and this was a pretty convincing advertisement.

There were a couple of weeping willows to my right: trees you normally see in parks and gardens. These two, no doubt the result of pure chance, give a slightly suburban look to the wild landscape. They were more forward than the more appropriate crack willows: the just-bursting buds were a strange shade of yellow. The dyke before me was claggy with an algal bloom: the other day I had noticed a moorhen walking on the stuff.

I sat for a while longer, looking at marsh harriers, the two familiar females. There was a grunt and a grumble from the heronry and, making my first movement for a while, I savoured a small smile.

> 🌿 This worry thing is a delicate balance. We need to be aware of the problems and support the righteous organizations. That doesn't mean there is no joy in what's left.

Thursday 26th

Sorry. Sorry, dear reader. Sorry, world, and perhaps a few apologies are due even to me. Another morning of computer rage had been followed by rage at the pointlessness of rage. Oh dear. And then came the shame – rather worse. Oh dear, oh dear. I took my seat in weariness and self-disgust, also in a cool breeze, under high cloud. Sorry, trees. Sorry, marsh. Sorry, birds. Of course I know these excesses are really about my father, but that knowledge is no help. Every day I have sat here in my seat, this peerless view before me, he has been in my thoughts, sometimes just unproductive worry, sometimes still less productive guilt, sometimes wishing I was sitting instead on a train to London, Covid restrictions a thing of the past, no fears of bringing the virus into his house – and sometimes finding a nice bird I could tell him about when we spoke that evening, or remembering a good story from the past, one we could work over together all over again ... like the night when, over-fortified by Glen Grant whisky, he had announced that he was a Christian Marxist. That brought me the tiniest possible smile. Good, better.

A little owl called; a robin and a great tit sang. A heron crossed the marsh, working hard as the wind got up. Then, instead of making the usual reared-back parachute descent with legs a-dangle, the bird tipped forward and made a beak-first dive into the dyke. Nice adjustment.

A Cetti gave a great burst of song and a few seconds later, there was another, fainter. Further away? One bird answering another? But I guessed not: this felt like a single bird pretending to be two. A rival would jump to the conclusion that there were two males already making claims on this desirable spot, and move on. It's called the Beau Geste stratagem.

A marsh harrier was gliding over the distant farmland. Or was it a buzzard? I am horrifically incompetent with computers; I also have many weaknesses as a birder, but I felt no urge to hurl my chair into the claggy dyke in rage at my uncertainty. And it *was* a marsh harrier, tail fanned in the soar.

I heard again the terrible roar of a vast plane, scanned the sky for a sight of the damnable thing, found a flying object high in the clouds, adjusted focus – and it was a harrier, tail now unfanned, making the perfect harrier silhouette. The discovery of this high, high harrier – 1,000 feet? More? – made me laugh out loud. You can call that healing.

> 🍃 A little reading around your wildlife experiences will produce all kinds of gems like the Beau Geste stratagem: the tactic of pretending that there are more defenders than there really are. Some good sources: *Birds Britannica*, by Mark Cocker and Richard Mabey, and *All the Birds of the Air*, by Francesca Greenoak.

Friday 27th

The wind had got up again and the roar filled the valley. At least it wasn't the plane, but it had brought rain and cold with it. Still, the robin was singing beside me. The sallows were now just about in leaf. I could hear the distant whiz of Cindy's power tools: she was making barn owls out of wood. The wind coughed and raised its voice while the raindrops made circles in the dyke as big as fried eggs. But the robin sang on.

That was it. Quiet sit. So greedily I went back for more later on, around five, as the light began to fade. By this time the rain had stopped and the wind relented a little. The Cetti sang out again, this time to my right. Another bird? Or more Beau Geste stuff? Keep listening.

A marsh harrier – two – no, three – all manoeuvring noncommittally over the river, forward and back, round and round. Was there a male there? Hard to tell. Keep watching.

And still the robin sang. A single curlew cried out, probably the same bird that has been here for a week or so. Fly north, young bird, fly north.

> Listen for the suddenness in the song of the Cetti in low, wet places: this is a bird that likes to make people jump.

Sunday 28th

The weather was warmer, the wind rather more fierce. A kestrel flew from the willow stump and a hare streaked across the marsh. I was largely deprived of a sense: the wind dominated the soundscape with whistles, groans and roars. Was that blackbird I could hear behind it? Or mistle thrush?

A mallard troika flew past, curving away: a piece of formation flying I had been aware of for the past two or three weeks. It involves three mallards in flight, usually the female leading, the next male close behind and a second male in the rear. They generally circle the marsh two or three times before the trailing male drops down. The rear male is chasing an interloping pair from the territory he shares with an unseen – and perhaps already incubating – female.

There are plenty of mallards out on the Flood, and plenty of troikas as a result. It seemed to be part of a daily routine rather than something that gets settled once and for all. As I watched, the rear male dropped out of sight and the remaining pair carried on in a long, fast curve.

A female harrier made a crossing and there was a perfunctory message from the Cetti. Still here.

> Information on the mallard troika (they don't call it that) from *BWP*, that is to say, *Birds of the Western Palearctic*. Every time you see a bird doing something you don't understand, you will find your answer in its inexhaustible pages.

Monday 29th

The wind was still brisk, but it was sunny and warm. As a chiffchaff raised his voice I made a bet with myself: today I would see the first butterfly of spring, or at least the first from my seat. A heron made a low, rather ponderous run along the dyke.

And then, do you know, I had just a wink of butterfly: a hint of fluttering black that was almost certainly the underwings of a peacock – the butterfly, not the bird. A mallard troika flew over. A queen bumblebee passed over my head, labouring as she went. It was all rather hectic. The pursuing male mallard abandoned his pursuit and headed back to the Flood.

A movement in the ash: in the still bare twigs a chiffchaff was hunting for tiny invertebrates with its sharp little beak; I could see the hint of yellow under its tail as it caught the light and once again marvelled at the marvellousness of the most ordinary birds.

Two notes, just the two, from the Cetti. A bumblebee, the same one or another, circled my seat. A marsh tit sneezed several times to my left. Somewhere in the distance a blackbird sang.

> The only bumblebees that survive the winter are queens. A queen wakes from hibernation (technically diapause) to brave the chills in her furry coats, seeking a nest-site for the eggs she carries within her, fertilized the previous year. Once she has raised a brood the young worker bees emerge and forage while she devotes all her time to egg-laying.

Tuesday 30th

The wind was now a light breeze, the sun was bright and warm. No bravery or self-delusion was required to accept this as spring. I could hear great tit, dunnock, wren and chiffchaff agreeing with me. For the first time I hadn't added to the garments I was wearing at my desk.

A harrier flew from the lone alder; another, apparently larger, followed her and the two began to spiral. Surely it had always been like this: warm, kind and buzzing. I would never need those additional layers again: from now on I would always be able to sit in the sun to the sound of birdsong. A blackbird sang, then a chaffinch.

A peacock butterfly banked sharply, showing me the four eyes of its four wings – and below it I could see a few lesser celandines; surely they weren't there yesterday. A hare fed on the open marsh; another butterfly flew over its head. Two more peacocks came together: there is a move in Chinese swordplay dramas when two fighters clash in a great clangour of steel and rise 20 feet up into the air while exchanging the most fearsome blows. That's roughly what the two butterflies did.

This was no day for the restful contemplation of nature and the thinking of deep thoughts: far too much going on. A drake mallard flew across; I heard the thin high note of a treecreeper in the ash. Soon I'd go indoors for a rest.

> Lesser celandines are great pioneers, blooming impossibly early, low to the ground like little yellow suns fallen to earth.

Wednesday 31st

Warm and content, I was watching a female harrier and listening to a blackbird in full song, sweet and melodious. A stock dove provided a rhythm section.

Then the harrier flew across the marsh more directly than is usual for harrierkind; unwaveringly, she flew straight at the lone alder. A kestrel – I had missed him, damn – flew out at some pace, proving yet again that birds are the best birdwatchers.

The kestrel flew off in the direction of the church, the harrier close behind, at least at first. She didn't have the straight-line speed to trouble a falcon, but that wasn't the point: she was most certainly in charge. Call that an investment in futures: this land is here to feed harriers, not kestrels.

> Blackbird song is as close to conventional human music – a real tune – as any birdsong you will ever hear.

APRIL

Thursday 1st

The temperature fell with a thump you could hear for miles. It was cloudy with a determined northeast wind in my face. The whole world seemed astonished by the change and the land before me was almost silent. It was like working for a bully, as I did for a while in my youth: he never forgot that a hint of kindliness would give proper savour to the next cruelty. I could just make out a hint of chiffchaff and a repeating phrase from a marsh tit, but there wasn't much conviction in it.

It was Maundy Thursday, Easter just ahead, the holiest time of the year, and in normal circumstances my father would be with us celebrating new life, renewal and hope, but this was the time of Covid. There's an undying schoolchild in every one of us: a shrill, bewildered person who still expects life to be fair.

I sat there thinking about a piece on anthropomorphism, reflecting on anthropomorphic images of non-human species in Beatrix Potter, Walt Disney and the Sistine Chapel: they're part of the way we think. A bumblebee was making rather desperate progress against the wind, not a good day when a queen has it all her own way. A heron rose up from the dyke, made a few dignified flaps and dropped down into a better position, knowing that anywhere out of the wind was a better position. Or was I being anthropomorphic?

> We can find some kind of understanding of other species in our imaginations, in the certainty that it can never be complete ... which is also true of other humans.

Friday 2nd

It was Good Friday: the second year running that we had to mark the day without my father. I wanted to have something good to tell him that evening, and I had a strong feeling that today would bring the arrival of an adult male marsh harrier.

I made a note of that in my book. But first came the male kestrel, initially perched in the lone alder and then hovering over the reeds: the place is only reserved for harriers when the harriers are there.

A ragged female harrier passed overhead at 300 feet, too high to bother with kestrels, feathers missing from wing and tail. Had she been in a barney? As I scanned the sky I found a small group of distant birds, and from the way they were flying, all swoops and curves, I knew they were either swallows or martins. Too far to be sure, but a modest bet had them as sand martins.

And there he was in all his splendour: argent, sable and chestnut, beautiful in the drama of his return: a male marsh harrier. There's always a satisfaction shot through with relief when a migrant returns: as if order was being restored, or confirmed. The wild beauty of this place was uncompromised. And here, after my earlier notebook entry, was a rare example of hope instantly fulfilled.

> Difference between genders – sexual dimorphism – is drastic in many species, like lions and harriers. It is more subtle in species like swans and humans; in others, like kingfishers and warblers, it is very hard to separate the sexes by just looking.

Sunday 4th

Easter Day offered a moment to sit out, a little later than usual. Every year since my mother died almost thirty years ago, we had spent Easter Day with my father, often in Cambridge, where, after a celebratory lunch, he would punt us all sedately along the Cam, delighting in his skill with this awkward craft. He did this every year until he was in his mid-eighties. Not this year. Not, as I think I said before, fair.

A female harrier crossed the marsh and in the background a blackbird sang with sustained commitment. It was sunny and cool, very nearly pass-the-Pimm's weather.

Very high, two birds of prey, one clearly a buzzard. I had the other for a kite, but it was too high to be certain. A well-fanged deer crossed the marsh just in front of me with the greatest nonchalance, quite unaware of my presence. The wind, blowing briskly from the north, helped to make me – well, I want to say invisible, because unsmellable doesn't have much force for us scent-deprived humans, and besides, it's mildly comic. I was unsensible, then: drawing no eye, making no sound, my scent blowing from marsh to house.

And there was the kite hammering towards me – no, it wasn't. It was the male harrier; of course it was. It had been a while since I had been on terms of daily familiarity with such a bird, and my mental template needed refurbishing. The pale leading edges of the wings were especially striking, with the light coming over my left shoulder.

As the harrier swerved right towards the church, I paused to assess my easy assumption that I would now be seeing male harriers almost every day. Every now and then you have to remind yourself that humility and gratitude are an important part of every privilege we enjoy.

> These mental templates are at the heart of good wildlifing; they develop with frustrating slowness across a lifetime of looking. I can tell a marsh harrier at a glance – well, often enough – but a top birder can also pick out a pallid harrier from any other species of harrier with the same certainty.

Monday 5th

The day had begun with horizontal snow. It was still intensely cold when I took my seat at about four, with a serious wind in my face, though an icy sun was shining. There were two swans in the dyke just before me. I had seen them before I sat down, and I moved with what I hoped was tact and discretion. It takes a lot to worry a swan; they just paddled a little further along the dyke, happy to be out of the wind in a place that allowed a little gentle snacking.

A male harrier sideslipped across the marsh, wings hunched in, swift and economical. He then turned at 90 degrees into the wind, stretched his wings out fully and stopped dead, as if he had an airbag. From there he tipped forward and dropped headlong out of sight.

Ten jackdaws flew over, playing in the wind like puppies. Two shelducks crossed, recrossed and re-recrossed, apparently reinforcing their togetherness. A lesser black-backed gull rose from the river and, with a series of slick, unostentatious manoeuvres, showed that when it comes to mastery of windy conditions, the long, slim, whangy wings of a gull are as good a design as you can get. Seabirds are best at wind: they have to be.

> In truth there is no best way to fly: it's about whatever works best for the lifestyle of the species. A good understanding of Darwin helps every observer to understand what's being observed.

Tuesday 6th

It didn't look or feel like proper snow. It was more like the polystyrene balls used for packing fragile items: I almost expected to see a man in the sky shovelling them down. The wind was brisk and in my face: snood weather. But there wasn't much for my exposed ears to hear: just the rooks in the rookery and the daft crowing of pheasants.

The snow stopped, the light grew a little brighter. A bird moved: it was a moment before I realized it was the male harrier. The capricious weather shifted and snow came more thickly than before. A Cetti shouted his defiance.

Then shockingly the sun was out and I was looking in vain for a snowbow. The marsh before me was shining and the swirling snow looked like broken shards of pure light. Then, slowly, the sky grew dark once again and the snow changed its nature, coming down in huge soft flakes.

And then it paused. The male harrier got up again and at once a female joined him. The male wig-wagged his wings extravagantly, the female circled him approvingly and then, after the briefest dance imaginable, they were gone. The wind shifted to the west and doubled in force.

> Courtship behaviour is deeply rewarding to every observer who keeps looking after diagnosing the species. Here the male bird was seeking to impress the female, showing off his beauty and his flying skills – incidentally impressing a human observer as well.

Wednesday 7th

No snow. It was even mild if you remembered the previous day: cloudy with a fair amount of blue. The two swans in the dyke showed no inclination to panic as I sat.

A female harrier was briefly in the air near the heronry: I always like the way they increase the dihedral to drop, raising their wings to lose a few feet under their own weight, like a shuttlecock. I had a brief glimpse of the male harrier: recognized him at once.

I could hear the swans feeding. After a while they drifted into sight: which was the cob, or male, and which the pen? Males have a more pronounced boss at the base of the beak, but I couldn't work out whose boss was bossier. Poor observation skills.

A bird flew overhead: concentrating on the swans and focused close for that reason, I was too slow to adjust for this new arrival – but it made a sound like a raven. That would have been a new bird for the marsh and a great excitement; I've been waiting for one for years. But I played the call back in my mind and it wasn't as raven-like as all that: just one of the sounds that rooks make. Better birders than me must live without such moments of excitement. Three buzzards were climbing in a stack. I checked, but none showed any sign of being a white-tailed eagle.

> Bill Oddie writes about the excitements of mistaken identity in his *Little Black Bird Book*, in a chapter called 'Brightening up a Dull Day'. You should never say 'it couldn't possibly be', because one day it will. White-tailed eagles have been seen over Norfolk: just not over this bit. Yet.

Thursday 8th

It was sunny with high clouds, the wind at my back still brisk. A chiffchaff was singing well and a mallard troika made a circle: is there no rest for these poor creatures? Two harriers appeared, the male flying a little higher than the female, both going through slick, assured manoeuvres.

My parents had a friend in the RAF who sometimes took us to Farnborough Air Show. I was reminded of this when the male kestrel came fizzing along with the wind behind him. He stopped to demonstrate first the power-hover and then the glide-hover, the one when he glides, wings still, at the exact same speed as the wind and so remains stationary relative to the ground. There seemed no intention of serious hunting: just showing the audience the range of his flying skills.

A great tit sang loudly from the ash. A long-tailed tit called: earlier I had seen two of them gathering the cobwebs they use to make their exquisite nests. The distant sallows were now a good sage green, from the catkins rather than the leaves, and I could see another clump of lesser celandines. I stood to leave and made a neat Ronaldo stepover to avoid an enormous turd. The swans have taken a fancy to this piece of ground.

> All right, the air-show bit was somewhat anthropomorphic, but it's possible the bird was practising: keeping his flying skills in good order.

Friday 9th

Curlews! As I stepped out there were twenty-three of them flying east. Were any of them thinking about staying here and breeding? It's happened before. The day was mild and still, overcast. A queen bumblebee buzzed round and round my chair. Movement over the river: after a bit of looking it was clear they were swallows, twenty or thirty of them a good mile off: the first unambiguous swallows of the year. Huzzah!

A buzzard soared over the marsh. There's sometimes a moment in a buzzard's soar when they thrust the leading edges of their wings beyond 180 degrees to make an angle of perhaps 160 degrees. It looked, to my human eyes, like the ecstasies of love: a willingness to embrace the entire world, angel-like.

The male harrier was perched on a distant post, occasionally making a small cry a little like a lapwing. And then a fragment of song and it was surely – well, almost surely – a blackcap. I could now count fifty swallows as they came a little closer. The Cetti sang out. And then, for just a few seconds, the blackcap sang properly: marvellous and melodious, one of the great singers of the English spring. I sat on, filled with thoughts that lie too deep for cheers.

> Blackcaps are your reward for getting a little deeper into birdsong. Some people say their song is better than a nightingale's; certainly they're a good deal easier to find, in urban parks and suburban gardens as well as tree-filled countryside.

Sunday 11th

The day was seriously chilly, but the sun came out and made everything all right. Then it went back in again. Nice cop, nasty cop. The sun lit up the male harrier with brief extravagance and then switched him off again. I put my gloves on.

I was deeply uneasy, horribly conflicted. Travel restrictions had been lifted, and I could go to London and visit my father. That was good. But it was also bad because I didn't fancy the tensions of travel or the risk of transporting Covid into either house. I was eager to see my father and worried that I might find him greatly diminished. These had been hard times for us all, but beyond question hardest for him.

A lone swallow flew over the marsh, the first close swallow of the year: let's hope that the summer is coming and will be a good one. God knows we deserve one. A flying swallow is a traditional tattoo for sailors: a hope not just for summer but for land: for home: for better times ahead.

A call from a distant Cetti – and then a sensation.

A sleek, muscular raptor flew over fast, flap-flap-glide, round wings, hint of bow on the trailing edge, hint of pallor between body and tail. It was a thing of startling power, speed, glamour, exoticism, beauty and excitement and I had it in sight for maybe five seconds. The wonder of it – and it was either a female sparrowhawk or – well, a female sparrowhawk is about the same size as a male goshawk, is it not?

I had a strong impression that this wasn't a sparrowhawk. I felt this in my water ... but that's not enough evidence to hang a man, is it? I've never seen a goshawk here before, and – but no, I can't claim it, I really can't. It was quite definitely a possible goshawk and quite possibly a probable goshawk. Over the heronry three more birds of prey, all harriers. Definitely. Any more out there?

> It pays to cherish people who know more about it than you do; after all, teaching enriches both pupil and teacher. My good friend Carl is ungrudging in his help and he agreed that goshawk was a decent possibility, adding 'your call'. Birding is all about moral character, then.

Tuesday 13th

London had been spooky and uninviting, my father a little feebler, a little more forgetful and it was very damn good to see him. Very damn troubling, too, but we drank and we talked and we laughed, sometimes as if troubles had never existed. Now I was stepping out to my seat again, taking a circuitous route along the dyke to avoid disturbing the swan who was lying on the grass about three feet behind my seat. I sat down with a shot-in-the-back prickle between my shoulders.

A Cetti sang; a female harrier dropped into the reeds with her undercarriage down, making a rather awkward shape. The male came to join her: domesticity in full swing, it seemed. When I stop seeing the females I'll know they're on eggs, leaving the male to do the foraging.

Strange sounds from the dyke: must be the second swan. A treecreeper called from above my head. The second swan was now looking up at me from the dyke: I looked at the bill but was still uncertain. Both bills seemed to me adequately bossed. Two males? The bird kept throwing meaningful glances in my direction, but they were probably aimed at the swan behind me.

A burst of piping: an oystercatcher was swooping down on the male harrier: they'll have a lot of work to do if they're to get any young off. A reed bunting flew over the marsh: not a usual bird to see here, and a pleasing one: it meant that the reeds were increasing and the marsh was becoming marshier.

I was back in my hut when I heard splashing from the dyke. The two swans, assuming a magnificent joint heraldic pose, were apparently, in the immortal worlds of William Blake, in lovely copulation bliss on bliss. Perhaps not two males.

> 🍃 Birds are about habitat. The reed bunting liked the greater reediness of this place. An understanding of what species you find in what habitat not only enriches your understanding of wildlife; it also helps you to identify them, on grounds of probability.

Wednesday 14th

It was sunny and breezy. A swan, apparently male, was sitting right by, though not actually in my chair. I approached in what I hoped was a non-confrontational manner, and after giving me a good old stare he moved about two feet. Up close the bird was almost incomprehensibly huge: an unbirdlike size, an unflyable weight. I said thank you and sat.

There was a good loud great tit and a rather tentative blackcap. I looked over my right shoulder: the swan was now walking away, picking his huge webbed feet up high like someone about to go snorkelling, achieving a dignity I have never managed in such circumstances. He then sat down again and proceeded to crop grass all around, quite unthreatened. It was the thank-you that did it.

A half-baked song from a goldfinch; a buzzard flew from one of the weeping willows, which were now a proper green. Three swallows appeared from behind, flying over my head and, using the wind without fuss, they cruised about the marsh, mostly gliding, occasionally jinking after an insect with a quick snap of the wings.

> 🍃 A mute swan is heavier than an Andean condor, a great white pelican and a sarus crane.

Thursday 15th

It was overcast, a chilly breeze in my face. The swans had avoided the wind by entering the dyke; as I arrived they looked up at me in a way that reminded me of a saying of Cindy's great-grandmother: 'I've got my north eye on you.' I spent a while looking at the bosses on those bills without any firm diagnosis of gender. At least one of us was confused.

As a goldfinch sang jauntily from the ash my eyes were attracted to a rapid movement in the sky: the male harrier was harrying a buzzard. Harriers are much more agile than buzzards, and, it seemed, more belligerent. Two little owls called suddenly, a sound like someone treading on the paw of a Yorkshire terrier.

The harrier vanished behind the lone sallow and I felt a touch of irritation: I had intended to have the damn thing down in the winter. But perhaps a bird would nest in it this spring. Sweet grapes.

A great black-backed gull flew over, the black mantle startlingly dark in the murky light. A buzzard pulled out of a soar, tucked his wings in tight, reducing their area by nearly half and dropped 200 feet in five seconds. It then extended the landing gear to brake and pulled out swaggeringly into straight and level flight. Applause?

> Great black-backed gulls are much bigger than lessers, but you can't always get an accurate idea of size, especially at distance. If the gull's back is charcoal grey, it's a lesser black-backed gull; if it's coal black, it's a great black-backed gull.

Friday 16th

Sometimes a wild day goes soaring beyond the commonplace in the most effortless manner, leaving you in a state of transcendent delight, while the world before you gets on with its business as if it's the most natural thing in the world, and for the best of reasons.

It was bright and sunny, a cold breeze occasionally gusting in my face. A female harrier was soaring in tight circles to the right of the heronry. A marsh tit called; another or the same harrier flew low over my head and moved east before turning and coming back. I could hear the male calling but I couldn't see him.

Then all at once I had him clear, at about 300 feet, and the moment I had him he began to dance: yes, here was the famous skydance of the harriers performed in the ballroom of the sky above my seat. He began with a series of wig-wags, showing first the top of his wings and then the underneath, one wig to each second.

I have suggested in semi-jest that a flying bird was looking for applause: no question at all about the fact that this was a dancer performing to an audience and hoping – perhaps even arrogantly expecting – approval. It was a dance to show off his beauty and his skill: his overall and unarguable mate-worthiness.

And then, having warmed up, he went into serious action. He started to loop the loop at high speed: and it was as if he was being whirled round on a piece of string: again, again, again, and as the huge unseen aircraft had filled the sky with its hellish din, so this dancing harrier filled the same sky with beauty and wonder. I sent him wave after wave of silent applause and I have no doubt that his target audience responded as well. How could she not?

> 🍃 It's the seen-it-all observer who misses stuff. It's good to maintain a certain level of naiveté: if you keep watching – rather than saying, oh yes, marsh harrier, obviously, without breaking stride – then you will be rewarded.

Saturday 17th

The swans, just a few yards away, paid me no mind as I sat. That was generous of them, because it was their garden now. They had taken charge. A little earlier one of them had hissed at me ferociously when I tried to enter my hut. Cindy was making an immense sculpture from mild steel, centred on a kind of co-existence tree. These swans were putting these generous sentiments to a searching examination.

It was cool and sunny. A male harrier made a long left-to-right glide; a couple of minutes later a male harrier was flying over the reeds near the heronry. Was this one a little darker? Were there two males out there?

A blackcap gave out a fragment of song, and then a Cetti.

> 🍃 I have spent a lot of time with horses, and also in the African bush, and both experiences teach you the importance of confident body language in trying circumstances. That's the way I was dealing with the swans. You can see this principle brought to its logical conclusion by checking out my old friend Manny on a YouTube video *Manny vs Elephant*.

Sunday 18th

I took my seat in mid-afternoon full of new information, and I must break the unities in order to pass it on. A couple of hours earlier, while paddling my kayak on the river, I had come across a group of harriers all in the air at once and counted five, two of them adult males. One of them was dancing, constantly rising and diving back down again. Between them these five birds would set the tone for the months to come.

Back at home the swans were grazing and glaring. They'd been having a difficult day: Cindy and Eddie were making a fence, weaving lengths of willow and hazel between uprights, banging them tight with mallets. The swans prefer their garden without intruders, or, for that matter, fences.

The nettles before me were getting thicker; there were two fat dandelions in flower. A white-tailed bumblebee orbited my chair and a Cetti shouted. A male harrier – I should say, one of the two male harriers – was cruising low along the river edge. We might get as many as three nests, one of the males servicing two.

A blackcap and a blackbird duetted: the two most melodious birds in the English soundscape. Were they doing so on purpose? They're not in competition for each other's mates or territories, but perhaps each was stimulated by the others fluting. After all, they're both musicians. Song thrushes, with their powerful repetitions, and nightingales, with their extravagance, are better at filling a landscape with music: but for sweet, sustained melody these two are perhaps the best of British.

A heron flew into the heronry carrying a stick bigger than itself, in the manner of an idiot dog. No exaggeration: the stick was sticking out beyond the trailing legs. It was big enough to be hammered into Cindy's fence.

> Wildlifing is not restricted to formal occasions with bins and notebook. On a kayak you are more or less sitting on the water and that gives you a privileged view of a different kind. In the winter I see kingfishers most days I paddle.

Monday 19th

Blindfolded you'd have known it was spring: a blackcap behind me was giving it everything. Blindfolded and earmuffed you'd have known it was spring: I could feel the sun's tentative warmth, and what's more the elusive scent of things growing tickled my nose.

And there was the blackcap in plain sight, dapper and jaunty, his black cap on quite straight. A typical blackcap phrase lasts about ten seconds, usually beginning with harsh and scratchy notes before it becomes more melodic and rises to a peak of flutey sumptuousness. This singer had the habit of starting with his head level, and then raising it gradually and with great theatricality until, at the conclusion of each phrase, his beak was pointing skywards.

It had been a small adventure to reach my seat: I took a roundabout route to avoid the nearer swan, from a mixture of courtesy and caution. As I sat there watching the bird cropping grass I wondered why the association between the two of them had not produced nests and eggs, for such things were now seriously overdue. I looked at the boss of the beak of the bird in sight and the penny dropped with a clang: they really were both males. What, you may ask, of that heraldic apparent copulation? What indeed?

I counted five exuberant dandelions at my feet; a honeybee touched down on one. There was some reassuring yammering from the heronry. The two swans were now swimming together in the dyke: surely I was right. What took me so long?

> Homosexuality has been recorded in many species in the animal kingdom; in his book *Biological Exuberance*, Bruce Bagemihl claims 450 – so we should never discount our own observations and experiences with the idea that 'it couldn't possibly have been ...'

Tuesday 20th

The blackcap had moved into the ash above my head and was singing well, enjoying the sun and the still air. There were more lesser celandines on the edge of the dyke. A swan walked up behind my chair and hissed, but I sat tight.

Your brain can play tricks, offering the wrong visual template in response to scanty information: for half a second I was watching an African fish eagle, a daily sight along the Luangwa River, though not the rivers of the Broads.

But it was a heron, of all things, pale head catching the light. There was no weight-bearing beam in his beak this time. A small white butterfly – that's the name, not a description – crossed the marsh. The Cetti sang out sharply.

A peacock – again, that's a butterfly, not a bird – was briefly in sight, so that's the first sit this year with two butterfly species. The blackcap struck up again; I could hear the swan cropping grass ferociously. The Cetti called again. I made a socially distanced semi-circle around the swan as I left.

> 🌿 'You've got good eyes.' Non-birders will sometimes say this to me as I name some reasonably distant species. But it's not my eyes: it's the collection of these templates I have collected. What I've actually got is a goodish brain – at least, when it comes to recognizing birds from quite a long way off.

Wednesday 21st

A significant date: not just Joseph's birthday but the anniversary of the day the first cuckoo of the year flew across his birthday party in the garden and cuckooed from the big oak.

This year we would have a small gathering to mark the day: two of Joseph's friends and takeaway pizzas: wild behaviour unthinkable a year earlier.

I found some time to sit out before the great event. Binoculars don't just bring birds closer, they also take you further away. They allow you to throw your soul a mile away and a couple of hundred feet up into the air. The shapes I could see were tiny, invisible with naked eyes, but still packed with meaning: two marsh harriers, yes, surely, and a little above them, a dozen swallows.

It was cloudy and the brisk northeast wind was keeping things pretty cool. It was also holding up the migrants: a headwind is the last thing you want on the last leg of a journey from sub-Saharan Africa, so many of them were waiting for the wind to drop before making the final push.

My bum remained on the seat but most of my being was now in the sky a mile off, and there were now four harriers, and after another moment, five: this last, certainly a male, folded his wings into a W, stressing the patterns of his wings, and glided steeply down into the reeds.

Now there were fifty swallows over the river, flying in those endless swallow curves. I remembered the passage in *Tristram Shandy* when Corporal Trim demonstrates the idea of freedom with a great flourish of his stick, and at this, the author, Lawrence Sterne, breaks free from his own prose to make a diagram of the stick's passage through the air, one that looks like a slow-flowing river with bends and meanders. The diagram is reproduced every day of the late spring and summer in the flight of every swallow that takes to the air.

> A long examination of a distant inaccessible place – the sky, the sea, the far side of the river – from a stationary vantage point enriches the observer in a strange and indefinable way, as well as bringing sights you would otherwise have missed.

Thursday 22nd

It felt just the same as it did yesterday, though no doubt there were a million subtle changes I was failing to notice: cool, sunny, breezy, the song of blackbird in my ear and the swans feeding in the dyke. Swans do a lot of feeding: they have a low protein diet of vegetation so, like horses, they need a lot of it. But unlike horses they fly, and excess baggage is always a problem, so they operate on the principle of an ultra-rapid through-put. The evidence of this was all over the garden.

Not that they were doing much flying. In winter swans are always commuting from one food source to another, but in spring they pair up and take season-long residence in a territory full of good food. These two had done that, but without the nest-building. I could hear the sound of their watery meal.

A female harrier went down in the reeds near the heronry and the blackcap sang in the ash, though without yesterday's commitment. To make up for this, a wren was giving it everything from the tangles to my left, really stressing that concluding trill. Video footage of wrens in such ecstasies shows a small body on the point of bursting.

The oaks, the ash, the alders, the crack willows were still

in bud. You'd think they'd be agog to eat up all this chilly sunlight, but they were biding their time. Overhead a crow was persecuting a buzzard with immense dedication.

> 🌿 Awareness of droppings is a useful wildlife skill. It works better with mammals than birds, but when you find the ground below a perch well-whitewashed, you may have found the favourite roost of a nice big bird, something to keep an eye on.

Friday 23rd

It was still cool and sunny but the wind had swung round to the east; one swan was under the ash and the other in the dyke. This was all important information to take in before sitting down. Two harriers were belting along the line of the river; the female vanished but the male paused for a moment and then stood on one wingtip to drop in a rapid spiral.

A buzzard crossed low over the marsh while the harriers had now climbed high, a reverse of their usual patterns. The blackcap set off into the most full-hearted song. Then a whitethroat joined in, first of the year from this spot. Once begun the song got stronger, filled with optimism: and then the bird was in sight, for they are the most visible of warblers, moving through the upright wands of the fallen willow: quiet colours with a neat white dicky.

A jackdaw flew across making an unusual call, repeating it four times as if giving pedantic expression to a password: *kick-off kick-off kick-off kick!* And then came one of those puzzles that are so pleasing to those who listen to birds: a bird that sounded like a goldfinch but clearly wasn't. He was out of sight, so it was all down to my ears (and brain). The same bird struck up in the manner of a chaffinch, but that wasn't quite right either, even though the timbre was spot on.

I allowed myself the luxury of an approving nod. This was an aberrant chaffinch, then. Would his unconventional approach find favour among the chaffinch hens? Or would they reject him in horror? I always wish such pioneers well; I always fear for them.

> Whitethroat song is traditionally described as scratchy; you most often hear them from the unkempt hedges and tangles of bramble. They sound like a blackcap played on an antique gramophone that needs a new needle.

Sunday 25th

I made it to my seat without disturbing the muntjac. The day was cold and cloudy, the northeast wind blowing from the deer to me. He was grazing away happily, about 20 yards off, occasionally raising his head with a jerk to scan for danger. After he had done so he usually groomed himself briefly around the withers, as if to convince himself there was no cause for alarm.

Two oystercatchers had very good cause for alarm: they were up in the air mobbing a male harrier who had cruised too close to their nest for their comfort. The harrier carried on as if butter wouldn't melt in his beak, while the oyster-catchers shouted abuse.

A goldfinch, a real one this time, started singing. The aberrant chaffinch had failed to include the fizzing and buzzing notes that punctuate proper goldfinch melodies. The male harrier rose suddenly from the reeds to the sound of a fanfare from the Cetti: I felt what my father would have called 'a lift of the heart'. I reminded myself to tell him about it later.

> As you watch wildlife in action you will notice odd bits of displacement activity: behaviour that helps an individual to cope with the stresses of daily life. You no doubt will have noticed yourself compulsively tidying a room when things get difficult.

Monday 26th

It was bright and sunny, the northeaster still coming in chilly gusts. The swans were in the dyke; a heron was labouring in the wind, neck extended, struggling to make a safe landing. Even a heron can be uncomfortably buoyant. And then – was it? Wait. Listen. And again, and yes, yes it was, faint and distant but unmistakably a cuckoo, almost a week late but unquestionably here.

One of the swans had clambered out of the dyke and was preening himself elaborately. Feather maintenance is of course essential for all flying birds, though this pair had been in the air for weeks living like feathered cows.

Six drake mallards flew over: a collection of disappointed

bachelors. And then behind them a falcon, and glory be, it wasn't a kestrel. The wings were too long and too sharp, and the dodging pattern of flight was all wrong: here was a hobby, a falcon that comes here for the summer to feed on dragonflies and swallows. The different nature of this different species was subtly but deeply thrilling.

There is anxiety in naming a bird as a hobby: you keep waiting for it to stop and hover and reveal itself as a kestrel and you as a bad birdwatcher. But not this time: the bird was flying in rapid darts and it then made a pass along the river revealing not a trace of orange on the mantle, only the steel grey you would hope for.

Two arrivals, two fine summer birds. Good day. Message to both: please stay, if you can.

> It's usually said that hobbies look like enormous swifts, and that's a pretty good rough guide to the shape.

Tuesday 27th

The wind had dropped, the sun was out and a blackbird was singing. The day had brought out the St Mark's flies. They are supposed to appear on St Mark's Day, April 25, but who's quibbling?

They hang in the air with their legs dangling, as if they were sitting on an invisible swing-chair. Their flight is uncertain, drifting slowly with sudden darts. Often you will see two flying as one: they have an old name of love-bugs – the fly doth lecher in my sight, as King Lear observed. There were distant swallows behind the near flies, confusing the brain with a complex perspective.

A male muntjac stepped out into the sun and his browny-black coat glowed. A cock pheasant walked in front of him: a vision of Asia, for without humans neither species would be any closer than the Black Sea. The willow buds were beginning to burst at last, though the ash above me was in no hurry.

High above I picked out a gliding sparrowhawk. It was no longer seeking to gain height so the tail was unfanned, straight and rather long, the edges of the wings making parallel lines at 90 degrees to the body. It was the silhouette of Thor's hammer. The bird, a male from the neatness of the shape, performed a 360 while continuing on the same straight line, like a skater.

A hare cantered easily just the far side of the dyke, one of three leverets I had seen mucking about earlier. It moved in a finicky manner, careful where it put its paws, stopping every now and then to scan the marsh for danger, but quite unaware of me. I remembered a fragment from *Ulysses:*

> *I am the boy*
> *That can enjoy*
> *Invisibility ...*

> 🍃 We don't all have expertise, but we all have curiosity. Though sometimes it needs waking up: I remember the day when I took the audacious step of looking up those flies in a book. Everything that lives is worth knowing about.

Wednesday 28th

A bird had roosted on my chair and left copious evidence of a comfortable night. Pheasant, from the quantity and quality. The wind had shifted back northeast: it stung my eyes and compromised my hearing, but there's no sense in complaining about the wind in Norfolk.

I was rattled all the same: the pages of my notebook separated and fluttered as I attempted to note the song of whitethroats. The wind had quietened down everything else.

A few St Mark's flies were braving the wind: not a good day for holding still in the air. A heron went labouring into the heronry. And then a flicker of movement and there were a dozen swifts, each one a classic sickle in the sky, and with them, as many house martins, turning to reveal their white bums. The swifts were gone in trice, riding the wind as I have ridden galloping horses, but the martins dropped a little lower and stayed. For a moment I could hear the sweet raspberries they were blowing at each other.

> That call of the house martins is a nice one to get lodged in the brain: *pr-prt*, says the nearest field guide; *trrrr-it*, says the second nearest. It sounds like a polite little fart, which neither says. (Raspberry is rhyming slang: raspberry tart, fart.)

Thursday 29th

A strange thing happens when you look at and for distant swallows, both with naked eyes and through the bins: it's sometimes hard to tell birds from floaters, the little shadows that drift across your vision and (usually, though a dramatic increase is a serious alarm signal) tell you that you're getting older.

The day was overcast with a gusty breeze. Two crows were harassing a buzzard: normally buzzards are rather disdainful of such attentions, but this one got angry: three times it rolled on its back in flight and, fully inverted, flashed its talons in the faces of its tormentors. They withdrew with an aggrieved air: that's not in the rules.

All along the line of the river I could make out hawking swallows among the floaters, their own flight patterns mimicking the curves of the river. A small white butterfly reminded me how slow this spring had been: a butterfly was still something of an event.

> This habit of attacking birds of prey is called mobbing even when it's a mob of one. When you hear a volley of alarm calls from small birds, it's worth checking out: they may have found a roosting bird of prey (or, less thrillingly, a cat). Crows and gulls are especially keen on mobbing.

Friday 30th

What are they doing? What are they thinking? What will they do next? Hard questions to ask even of our fellow humans: harder still to ask across the division of species.

There had been some rain and it was still overcast. I knew

what the song thrush was doing: routine maintenance of his territory, moving from one songpost to the next, a good loud burst of song from each: the big oak in the garden, the willow on the right, the willow to the left.

But the marsh harriers had me baffled. The male was up and down, up and down. He made a low, belly-tickling run over the reeds with a female, and then was gone. And then up and then down again, another half-dozen times.

Then he was up once again and this time the female flew straight at him; they pirouetted around each other just the once and then separated. The male flew on, over the open ground, so the oystercatchers were up at once, harrying (or mobbing) the harrier. After a while he drifted back: the female was in the air with him and they dropped to the ground together. This time they stayed down. Perhaps it was love.

> Sorry, that should be reinforcement of the pair-bond. A male harrier will pass food to the female in the air; it would no doubt be utterly wrong to compare this to a gift of flowers on Valentine's Day.

MAY

Sunday 2nd

Where were the cuckoos? At this time of year the valley is normally shaking to the sound of cuckoos, but I had heard only a few distant whispers. A cuckoo called PJ, radio-tagged by the British Trust for Ornithology, was back and cuckooing away in Suffolk. Why wasn't my cuckoo doing the same? The ecological anxieties we live with from day to day seemed to merge with the oppressions of Covid and daily worries about my father.

The buds had burst on the crack willows and they glowed with a very young-looking yellowy-green: setting off the three colours of the male harrier as he passed in front of them. There was a great volley of mad barking from the caged dogs further along the valley; I knew how they felt.

A Cetti called from an unaccustomed place to my right. And then I heard a broubrou. This is an African shrike species; the name is onomatopoeic and I know the call well, but not from Norfolk marshes. So what was going on?

Obligingly, the bird did it again: a soft cooing trill, and I had it. Song thrush. The bird was not going full on with his usual repetitions: he was trying a new idea to see if it worked: a musician at practice, in the manner of my older son Joseph with a guitar. A blackbird sang briefly and beautifully.

But I was still in great need of a cuckoo.

> The BTO is a good organization, one that does the hard work of organizing surveys and collating the numbers. Their website tells you how any species is doing; you can also send them your own records.

Tuesday 3rd

Where was spring? The west wind was not far off gale force, bringing cold rain. A lone St Mark's fly staggered in a gust and struggled to hold its position before being wafted away.

Half a dozen house martins flew along the river. And then

a little moment of perfection: sharing the same piece of sky, a marsh harrier, male from the slimness, and a single swift, both making utterly different and utterly characteristic flight silhouettes. But why was the swift alone?

A couple of weeks ago the year had advanced so far that I dared to hope I would no longer look eccentric when I sat out. Now I looked quite mad, rain was dripping from my hat and searching for leaks in my weatherproofs.

A crow fell out of the sky, dropping 200 feet as if it had been shot before picking up the wind and making a violent change of direction. And then I solved the lone swift problem: it wasn't a nearby swift, it was a distant hobby, for there was the bird again, closer and this time unmistakable: as fast and elegant as Bond's Aston Martin.

The song thrush struck up full and loud, from the same spot as the previous day. He went through a long series of variations, repeating each one with an air of triumph – and eventually got to his broubrou verse. I wanted to applaud as jazz enthusiasts applaud in mid-solo.

> Distance is always a problem when you have nothing to compare. It's the lesson Father Ted gave Dougal, comparing cows from a children's farm set with real ones in the field: 'Small ... *far away*.'

Wednesday 4th

Well, it still didn't feel like spring, even though it wasn't raining and the wind was less fierce. But there was spring my ears: praise the Lord, a distant cuckoo. Not exactly a bravura performance, but doing the best he could in trying circumstances.

There was even a little sun. The St Mark's flies were making constant Dougal problems as I scanned the sky: is that a fly in my face or a distant eagle?

A muntjac strolled at his ease across the marsh, pausing to eat every few steps. A few more calls from the cuckoo: and a blackcap and a whitethroat joined in to show that our flag – the flag of the spring itself – was still gallantly streaming.

I could hear soft farting sounds from above: three house martins. I radioed to them: we've got room for you here: there are ready-made martin nests attached to the barn. You'll love them. They flew on, blowing me a raspberry as they went.

> Bird counts (at least those made by eye) tend to underestimate the true numbers, as we have seen: but with swallows, martins and swifts there is an opposite tendency: the birds are so busy in the air you are likely to count each one a couple of times or more.

Thursday 5th

The struggle for existence is very clear on cold days in the heart of winter, when every day you live through is a triumph. But I could feel it now in this slow, late and disturbingly cold spring: how could you raise a nestful when the cold weather was killing your insect prey?

A chilly breeze was blowing from the west, but a blackcap

was singing and a swift – not a hobby – was flying in darts and jinks, picking insects – a few about, then – out of the sky. A blackbird struck up as if he intended to sing for ever: finding the relaxed all-day rhythm of a horse in the Ascot Gold Cup. A green woodpecker called. A male harrier flew ten feet above my head and wheeled right, in the direction of the church. A whitethroat and a blackcap struck up together and a wren trilled close by. Busy busy busy, as we Bokononists say.

> The uncomfortable fact is that all seasons are hard and that existence is a struggle for everyone. Without struggle, evolution doesn't happen, because evolution is about unequal survival rate. Two excellent biographies of Charles Darwin helped me to grasp this bleak but essential principle: one by Adrian Desmond and James Moore, another by John Bowlby.

Friday 6th

The previous evening a storm had turned the place white: opaque balls of hail too hard to be snow. It pelted down, bouncing from the hard surfaces, leaving behind a January landscape and a world of struggle.

Spring was cut off in mid-flow. No Cetti called. I couldn't hear a cuckoo. Would there be butterflies ever again? Still, the nettles were growing well; the exposed earth was now all covered.

A blackcap began a rich, fluid song and a green

woodpecker did the mad villain's laugh. A heron, flying over at 20 feet, untucked the long neck and twisted it hard to the left, as if it were a length of hose, and spilled from the sky as water spills from an upturned bucket.

A whitethroat appeared on the bank in front of me, busy among the stems of last year's cow parsley. A male kestrel and a male harrier briefly shared the sky, making their eloquent shapes, and a hare moved slowly away from me, pausing to nibble, and then holding still, looking, listening.

> It's not always easy to tell a hare from rabbit; hares are bigger but you run into the Dougal Principle. The difference is clear when they move: a rabbit scuttles; a hare cranks itself up on long-levered limbs and gallops like a little deer.

Sunday 9th

I rejoiced, for the world was full of flies. The weather had changed, warm and sunny though still windy, and it had brought out a great crowd or cloud of St Mark's flies. The love-flies were all over the marsh, apparently constrained by a glass ceiling at about ten feet: uncountable hordes seeking love and the chance to make more flies. It was the day we had all been waiting for.

A blackcap began to sing and it was the best blackcap of the year: a melody so rich and sustained that another blackcap was forced to answer and each lifted the other to new heights of performance.

A female harrier, huge and dominant, made a slow pass from left to right in an exaggerated high-winged glide. A heron was also gliding, travelling over the dyke, so low I could see only the tip of the arch – the keystone – of each wing. A whitethroat rose from cover to perform a classic songflight: rising about six feet and singing at the top of his voice, yo-yoing in an ecstasy of song. The heron got up from the dyke and moved to a new position 50 yards further on, as a gambler out of luck changes to a new seat.

> St Mark's flies operate the strategy of superabundance: each one is vulnerable, hanging in the air, but when there are so many the predators will never manage to kill the lot: the most effective and the luckiest individuals will survive and become ancestors.

Monday 10th

Looking for St Mark's flies was like having an optical illusion explained to you: the more you look the more you saw. They hung still in the face of a tough wind from the southwest. It was sunny, even warm when the wind took a break.

The air may not have been still but it was full of song: blackbird, chiffchaff, great tit, goldfinch. Serious overnight rain had brought a feeling of change: perhaps even a reclaiming of this lost spring. A handful of swallows flew along the river: I wished they could leave trails of light, like sparklers, mapping their movements.

Then – huzzah! – a butterfly, a small white. The wind was

buffeting and the place still felt uneasy but the blackbird's song was full of calm: both blackbird and flies had found an extraordinary stillness on a turbulent day.

A magpie flew across the marsh like a paper aeroplane across a classroom. The puffy clouds had lined up across the valley in rows, as you sometimes see in Dutch landscape paintings. Is that all about the flat land beneath them? A heron flew past, unaware of me ten feet below, the sun bringing out the colour of his legs, a van Gogh yellow.

The buds had burst on the distant oaks. I longed for this to be a watershed day: the day from which High Spring would burst in all its glory, as if the days of struggle were over forever.

> These cotton wool clouds are cumulus, the ones that, as Hamlet said, look like whales and weasels. A passing acquaintance with clouds and their meanings is another small step in making yourself a little wilder.

Tuesday 11th

It was warm in the brief moments between gusts, and certainly sunny. The blackbird was singing, there were plenty of St Mark's flies about and a small white butterfly flickered past.

A female harrier rose, made a strong, high-winged glide above the reeds and went down again. It was the third time she had done so: no longer confined to her nest and its eggs, she was now hunting to provide for her chicks. A kestrel rose behind her and held a brief high hover: two birds of prey in their most characteristic hunting attitudes.

Two crows flew by making throaty noises at each other. I thought about the silence of the Cetti and the cuckoo: the more effective you find the consolations of nature, the more you will need them.

So right on cue I picked up a distant sedge warbler: they were back on the marsh and making the best of things. I realized with a start that it had been weeks since I had noted my progress in sitting meditation. Perhaps I had reached a higher plane of existence – or perhaps I had reached spring.

> Sedge warblers are wet country birds. They sing from tangles and their song, not over melodious, is complex and tangled: so much so that it's said that a bird never sings the same song twice.

Wednesday 12th

The sun was on my back, the clouds stretched out before me and the air was fairly still. An orange-tip butterfly came hurrying past; a great tit was trying out a complicated eight-note phrase. The previous evening I had heard a cuckoo call several times. Against the odds, spring was re-establishing its grip.

A sedge warbler was singing faintly, a small white flew by and a whitethroat was calling from the fallen willow. A peacock – third butterfly species in fifteen minutes – flew near my feet and tilted its wings: they caught the sun and all four eyes lit up: one eye on each wing, not for seeing but for fooling potential predators by pretending to be an owl.

When surprised in the dark places where they rest, they will rub their wings together to produce an owl-like hiss.

A male harrier made a low pass. Oystercatchers called from the far meadow, which was now full of red cattle... and I remembered Sebastian Coe, winning his second gold medal at the Olympic Games of 1984 against all the odds. Spring, like Coe, was turning in fury on a legion of doubters: 'Who says I'm fucking finished?'

> Male orange-tips are white butterflies with prominent orange tips to their wings; the females are monochrome. The males are very visible in the High Spring, always in a desperate hurry as they move from place to place in search of females, perpetually late for a very important date.

Thursday 13th

OK, bloody nature, cheer me up: that's your job, isn't it? I had all sort of aches from over-paddling the kayak the previous day. I had spent hours on a finicky job of captioning and then my computer banned me from the internet. Now I was at last sitting outside and the east wind was like a knife.

I could make out a couple of swifts and there were martins over the river. The great tit in the ash was trying a new variation, seven notes arranged in an unusual pattern. The cleavers and cow parsley were beginning to green up the far bank. And it was cold, bloody hell it was cold, let's forget this spring and wait for the next. But spring is like

Bob Dylan: the only thing that I knew how to do was to keep on keepin' on.

I remembered the Arcadian spring of last year, the first Covid Spring, the great season of consolation in the first tranche of hard times. It had been sweet and kind and filled with butterflies, like spring in a fairy-tale:

> *And when he shakes his mane*
> *We shall have spring again.*

The prophecy that foretells the coming of Aslan and the healing of harms in *The Lion, the Witch and the Wardrobe*.

There had been a strange joy in that time, and it had helped us all to deal with the bewilderment and horror of the first Lockdown Spring. But this second Covid Spring was hard and bitter: one to endure rather than enjoy. A deer was feeding on the lone sallow, creating a neat horizontal browse-line. A swan laboured into the air and was crossed by a swift: the most aerial of all birds and the world's second heaviest flying bird sharing the same piece of sky.

I found a few more feeding swifts, and as I looked three of them broke away from the rest and came hammering towards me, things other than food on their minds. Was this a swift troika? Or were they just young birds getting above themselves? Why not? – they were above almost everything else.

> Wise people keep telling us to go into nature to seek wellbeing. My experience is that if you go into nature to seek nature, you are far more likely to find a spot of wellbeing.

Friday 14th

It sounded like May. The day was overcast and cool but I could hear blackbird, whitethroat, chiffchaff, long-tailed tit, rooks and a blackcap all shouting at once – and yes, a cuckoo from the distant right. A monster growled from the alder carr on the left: the heronry was sounding wonderfully unbirdlike.

The buds had burst on the lone alder, the two carrs nearby were now fuzzed with green; the ash was the only holdout. A cuckoo called again, but this time from dead ahead, in the willows on the far side of the river.

Two blackbirds sang antiphonally, then two chiffchaffs: the fraught competition of Maytime was all around. I tried to keep my eyes on the willows, hoping for a glimpse of the cuckoo. A crow performed a stall turn, a half-cartwheel in the sky followed by a sharp descent.

Now the cuckoo was calling from the heronry, and I was cross with myself for missing its flight. But a few minutes later he was calling from the willow again. There were two possibilities. One: given that cuckoos can travel briskly and I am a bad birdwatcher, it was the same bird. Two: given that cuckoos are pretty obvious birds and I have occasional moments of birding competence, there were two cuckoos.

> Cuckoos select a series of prominent perches to sing from. These are technically known as stud-posts. The aim is to attract a female from a distance: that's why the call is simple, utterly distinctive and far-carrying.

Sunday 16th

Half the sky was palest blue, the other half covered in blue-black clouds that looked ready to burst at any moment. I could hear half a dozen species of birds singing away; a rook sounded like a boy scout at his first bugle lesson.

A male harrier got up with a flourish; a blackcap was performing at his best. A cuckoo – should I say one of the cuckoos? – was calling from a distance, mocking my earlier pessimism. There were still a few St Mark's flies.

Pleasantly weary, I leant back in my chair and felt the canvas yield accommodatingly. I shall do a follow-up project next year, recording only observations made from a hammock. I would do so in all weathers, eyes determinedly closed, relying only on information from ears, nose and skin.

As I pleased myself with this fancy I caught a tiny flash of pale blue in the tail of my eye. I sat up straight and looked for its source. Hard to find anything ... but there, it happened again, two or three dozen small flies, each wing twice as wide as the body was long. The wings were transparent but shiny: the flash of blue was the reflection of the sky.

I followed the flight of a common gull and found a buzzard; I followed the flight of the buzzard and found a hobby. I lowered the bins in satisfaction, caught yet another small movement and raised them again, to find a male harrier. This spring was not so bad now.

> 🍃 The hammock-book fantasy will remain just that: but immersion in nature is increasingly a process of bringing in less-used areas of sensory perception: hearing most obviously, but also peripheral vision, something we use less and less because we use screens so much.

Monday 17th

Matt Shardlow, chief exec of the invertebrate charity Buglife, knows his stuff all right, and he reckoned the flies I had seen the other day were probably mayflies from the genus *Caenis*. I looked them up and he was spot on.

I took my seat, foggy after my second Covid jab. The sky was still a rum mixture of very pale sky and very dark clouds, and I counted half a dozen species in song. One was a robin: working on a second brood already, I guessed. There was a hissing and crackling from the unfallen willow to my left: there was a big domed magpie nest in its heights and the inmates were now big enough to make a proper din. I tried to catch a glimpse of them and noticed that the buds in the ash had burst at last. Then a movement in my peripheral vision and I turned to find two house sparrows, unusual at this end of the garden. I picked out a very high marsh harrier, tiny but very clear in that this-is-me dihedral: so remote, yet so clearly itself.

And then an event. A group of shelducks with a few gulls was passing to and fro: and there was a buzzard in the middle getting a hard time. The shelducks made perhaps twenty attacks, the buzzard far too low for its own comfort, flapping hard to get clear. As it finally managed it, I heard two simultaneous roarings: the heronry and a deer were both in good voice. It sounded like a singing competition between two dragons.

> There is often good stuff in scientific names. Caenis was a beautiful woman raped by the sea god Poseidon, who then offered her one wish. She chose to become a man, so she would never suffer such a terrible thing again. This wish Poseidon granted, adding an unbreakable skin as a bonus. As a man, now Caeneus, he met his end in a ferocious battle against the centaurs; it's all in Ovid's *Metamorphoses*, one of the great natural history books, if a little unreliable in places.

Tuesday 18th

This is a book about stillness, but I'm mostly recording movement. As I sat in a heavy shower, I wondered about what happens before and after the movement. A blackcap and a chiffchaff sang out above the sound of the rain on the new leaves and my hat. Four shovelers flew over: lovely weather for shovelers.

In the cold and brutal rain earlier in the year I had wondered about survival; but this Maytime rain was about growth: the wetness of the rain and the greenness of the land offered a chance to hear the trees growing.

On the far side of the river there were a few May trees in

blossom. The sky gave a long, rather contemplative rumble, like contented digestion after a good meal: thunder as a peaceful accompaniment to the life-giving rain. A lone swift went by, black sickle against a black sky, as a panther (more accurately a melanistic leopard) has black spots on a black coat.

A zag of lightning cut the sky in half and did so horizontally, a brief and startling effect. There was a long wait for thunder; in the pause the rain grew gentler. A few minutes later the sun made an altogether unexpected appearance, throwing my shadow onto the grass, while a swan scrambled out of the dyke and then, as if to make up for this inelegance, spread his wings and gave a triple flap. A small white butterfly flew past him: two quite different ideas about how to be a white flying machine.

> Melanism – being much darker than the norm – is common enough. You occasionally see black bunnies; in Hertfordshire there's a population of grey squirrels that are actually black; there's a pub called the Black Squirrel in Letchworth. It's a useful reminder that every member of a species is also an individual.

Wednesday 19th

I didn't walk to my seat; I tottered. It was the combination of the Covid jab and an out-of-the-blue commission, *copy right now, please*. I obliged, from motives you can probably guess, and took my seat feeling like a hero who had hiked to the North Pole and back.

A deer observed my totter and took off, jinking sharply to the left after a few strides to put off any pursuit. It's not the straight-line speed that's so remarkable: it's the agility and the ability to balance at speed.

I could have done with a hammock rather than a chair today. I closed my eyes to listen to a singing whitethroat. A blackcap was making his contact call. The weather was still unsettled. There had been a single quite tremendous thunderclap directly overhead a little earlier; it sounded as if the house had fallen down. Now the sun was dodging in and out of cumulus clouds.

A buzzard cruised over at 30 feet, finding just a hint of lift from a breath of rising air, twisting neatly to catch it and gain ten feet in a flap ... and then it all went wrong. With sharp calls of excitement a female harrier with lowered talons came down on the buzzard, and made four attacks in quick succession. The buzzard moved on up the valley, the harrier continuing her assault, but now with the undercarriage raised.

She then turned back, the buzzard gone, and made three yelps. I thought this was a call of triumph, but no, she had sighted another buzzard. She went for this one as well, and was joined by a crow, a bird always happy to turn a kerfuffle into a first-class row. The second buzzard left in a hurry, or as much of a hurry as a big, glidy bird can manage.

> It's useful to learn those contact calls as well as the song, for the purposes of identification and for the deeper pleasure of knowing what's going on all around you. The blackcap's contact call sounds like two stones being tapped sharply together.

Thursday 20th

A muntjac and one of the swans both saw me as I approached my seat, so I tried to make my body language meek. The muntjac gave up staring after a minute or so and got back to eating; the swan came bulling straight towards me. Perhaps I'd overdone the meekness. But he merely slithered into the dyke to feed. It was overcast, cool and windy, but the blackcap was in good voice above me. The muntjac had only one antler.

Suddenly there were two dozen martins over the river and then, just as suddenly, there weren't: a brief feed and move on, the current supply exhausted. A shoveler flew across the marsh: so many shovelers so suddenly. Subtle, hard-to-explain changes occur in any environment you watch closely.

Was this the slowest spring for butterflies I had ever known? Last year the sun had shone every day and butterflies flittered and fluttered in its warmth. In that uneasy time it was easy for so many non-human creatures, but this difficult and troubling spring told a truer story: all the glories of nature are the result of struggle.

I had at least reached a working hypothesis about the singing Cettis: there had been one, and now there was none. Had there been two, there might have been one left, but the hard weather of recent days had wiped out the only one. I had heard other Cettis along the river from my kayak: perhaps they would be back here next spring. If the weather was a little kinder.

> Antelopes keep their antlers for a lifetime. Muntjacs are deer, so they shed their antlers every year, usually in May; this one was halfway through the job.

Friday 21st

It's hard to step out into hard weather, but once you've stepped it gets easier. I was back in full gear and trying to avoid pejorative or anthropocentric terms for the conditions, but I failed. Once again I fell back on my mother's terminology: it was vile.

Chaffinch, blue tit and blackbird sang on through the vileness, rain and vigorous wind. But there were clearly good numbers of insects over the river: getting on for 100 martins were hawking for them, and as I looked higher I found more. It was a sudden vista of bioabundance: and these days I value such sights more than any passing rarity.

A breakout party of martins came over to the marsh, giving me a proper appreciation of their speed: you need pinpoint accuracy to catch a flying insect in your beak, and if you're going to take it by surprise you need to be seriously fast. Here was agility at speed, a little like the fleeing deer a few days back, but they were chasing, not fleeing.

Four jackdaws crossed the marsh looking like black bin-liners blown by the wind. A bumblebee bumbled onto the marsh: I wonder how much killing this bitter day would do. Then I realized that the blackbird hadn't stopped singing even for a second. Shut your eyes and it was a lovely day. Almost.

> The decline of insects is an increasingly desperate problem and insect-eaters like martins are declining as a result. It's important to support organizations like Buglife, and for people with gardens to encourage insects instead of killing them.

Sunday 23rd

Today was Eddie's twentieth birthday. His friend Callum came to visit, and, for reasons best known to himself, arrived dressed from head to foot as Captain America. For the first time in eighteen months these two great buddies were able to hug: the best present of them all.

The bitter wind had split one of the big willows in half – vertically. It looked like a single blow of an immeasurable axe. The crack willow had cracked: one half was still standing, the other had rocked back across the marsh. Leave it there, then.

The wind had dropped a fair bit, though it was still breezy. There was intermittent sun and a robin was singing sweetly. The alders were in full leaf; the ash above was almost there too. The arc of my view had contracted by ten degrees: the blackthorn before me was also in leaf. Should I move my chair to counteract this, or would that be a breach of faith? And then against the silver and black of the clouds, a brief glimpse of beauty with a silver and black bird. It was only – only? – a herring gull, but it caught me off-guard with a sudden shaft of joy.

A hint of movement on the ground, or just above it: two

ears, gold and black-tipped, appearing and disappearing. A hare stopped, moved, stopped – and suddenly it was running. I looked for a predator and realized that the hare was not running away: it was running towards. There was another hare. They came together and were instantly lost in the vegetation: which I now realized was remarkably high for this time of year. The plants weren't finding the rain so vile.

An orange-tip butterfly appeared: in this weird May it was a rich and rare thing. It travelled forward and back, and then flew two feet above my knees. I looked up and saw a buzzard; a swift shot underneath it. They are both great fliers, buzzards on broad wings, swifts on ultra-slim wings. It's about weight: a buzzard can weigh 1.4 kilos, a swift around 40 grams: thirty-five times as heavy.

Over the river there was a party of swifts, at least 100, probably a good few more. I hadn't seen so many there for years: is that because I've been looking more or because there are more swifts? Certainly this moment of bioabundance was down to a superabundance of insects, so there were all kinds of reasons for cheering.

> 🍃 Dead wood is good. Land-owners should note that it's always good to leave a tree where it has fallen, if you can do so safely, for a fall is only the next stage of life. A fallen willow usually keeps going from a prone position – a Suffolk farmer told me 'you can't kill a willa' – and even if it doesn't, rotting wood is vital habitat for many creatures, mostly insects. Even the tiniest garden should have a log-pile.

Monday 24th

A cold southwest wind, a skyful of black clouds, a break in the rain and a migraine. They don't come often, thank God, but they're somewhat unamusing. After I had gone through the first frenzies and the fractured vision stage I stepped out: a swan was between me and my seat. He hissed as I passed. It began to rain again.

A wren struck up with an extra-long trill at the end: wrens always sound defiant. Two shovelers flew over, their beaks like carnival masks. A blackbird, a blackcap and a robin sang together: a sweetness competition. Or perhaps they were summoning the sun: if so, it worked, lighting up the fat raindrops on the ground at my feet, making them shine like objects of immense value. There was a *plop* in the dyke. Water vole? Kingfisher? Frog? Fish? Windblown twig? I'll never know.

A distant harrier, male, once again that perfect silhouette. A swift flew over my head and belted off in front of me, no higher than 20 feet. It made a long glide: and I noticed that its wings were tilted *downwards*; exact opposite to the harrier.

A male blackbird perched in the ash: through the bins I could see raindrops shining on the shining black feathers. The migraine wasn't exactly cured, but it was easier to deal with.

> The swift's wings were in an anhedral. This too has its advantages: what you lose in stability you gain in manoeuvrability. You can find the same configuration in fighter-planes: swifts use it to intercept insects in flight.

Tuesday 25th

Another dark day, another migraine, and this second one was shocking, unprecedented, really rather alarming. I stepped out in search of relief. A blackbird sang his most soothing song, the sky got a little lighter – and then instant dusk. The rain fell in huge drops: it was impossible to think about anything except rain. I realized that I was bracing my body against it: do the poor sheltering beasts of the marsh do the same? I looked closely at a nearby deer, but I couldn't tell. The rain had no difficulty in finding the small gap between neck and collar.

A moorhen swam along the dyke, and the rain slackened and then stopped. This pause was welcomed by a chiffchaff. A blackbird sang up and a blackcap joined in: they were attempting a cure for migraine, no doubt. Certainly I appreciated the effort.

Then two swifts flew over my head: so different from all other birds that they made me gasp:

> *The swift with wings and tail as sharp and narrow*
> *As if the bow had flown off with the arrow.*

There was hectic movement in the fallen willows: three small birds, probably tits and probably brand-new ones, just fledged and trying out the world. It gets warmer than this, little birds.

There were now a dozen or so swifts over the marsh, making a Catherine wheel of themselves. A hobby appeared briefly, looking rather clunky among the swifts.

> 🍃 If I had put my hood up the rain would have drowned out the birdsong. The poem is by Edward Thomas.

Wednesday 26th

It was now clear that these migraines were a sideswipe from the vaccine, and a doctor's appointment was arranged. Before I set off I had a moment to take my headache down to the marsh, beneath a sky as overcast as my mood. (I should perhaps add here that I was and am very grateful for the vaccination programme, despite the inconveniences.)

But a blackcap was singing and there was a party of swifts over the river. Paul Theroux begins *The Great Railway Bazaar*: 'I have seldom heard a train go by and not wished I was on it.' Well, I have seldom seen a swift go by and not wished to be it. And was that a willow warbler, a tiny distant trace? I couldn't confirm it, but I know willow warblers.

Swifts seem to double their speed at will, in a single instant and while maintaining a straight and level glide. I looked at bird after bird, trying without success to detect a sudden whirring of the wings as they accelerated. A sailor in a yacht can trim the sails by tightening or loosening a rope, bringing about a sudden increase in speed ... perhaps the swifts were deliberately flying with ill-trimmed wings, flying with the handbrake on, in order to assess what lay ahead – prey, rivals, obstacles, companions, potential mates – and then adjusting the wings for maximum efficiency and a consequent blast of speed.

The swifts were again flying with martins: it was easy

to tell them apart even when I couldn't make out plumage details, even when I couldn't get a clear silhouette. Both have swept-back wings and notched tails, but martins have more jinks and fewer and shorter glides, while swifts have a good deal more flat-out speed. They accelerate out of turns and into dives, sometimes with a sudden whirring of the wings; but they also do that change of pace on the glide by using the wind as no other living thing quite can.

> Once a swift has fledged it won't alight on anything until it nests: and that can be three years. They feed and sleep on the wing and even mate on the wing: so there's a thought for your next incarnation.

Thursday 27th

It was still overcast and cool. I was still pretty overcast myself, though not at my coolest. I was planning to take my headache to London, pleased to be seeing my father, troubled that I had to cross London to do so. I had established a good Tube-avoidance route on previous visits: walk from Liverpool Street to Waterloo, crossing the river at the Millennium Bridge and following the South Bank from the Tate Modern to the Royal Festival Hall.

But there was at least time to sit for a while before heading for the station. I found nothing spectacular, only a pleasing Bokononist busyness: chiffchaff, corvids, great tit, a buzzard. A brief flash as a jay passed. The vegetation had taken an extra leap in the past twenty-four hours, especially the nettles. Across the river the blossom had the landscape almost frothing. I had a just-in-case waterproof bedside me: a small brown beetle landed on it and stayed, borrowing some of the warmth and energy stored in its blackness.

There was a small flock of swifts over the river: I was struck again by the amount of time they spent gliding. Their way of life – forever on the wing – is more economical than it looks. A male harrier crossed at 200 feet, and then altered his wings from a V to a W to make a long, shallow dive.

But I was thinking about swifts and the aerial life. For them the ground is another country: I played with the idea that swifts envy my sitting as I envy their flying. Even with the long glides and economically brief wing-whirs, it takes a lot of food to keep them aloft. The air above the river had been feeding their flight muscles for the last couple of weeks.

> Swifts are (usually) classified in the same order as hummingbirds, Apodiformes. All members all have long wings and short, stout bones in the upper arm. Swifts are the most aerial of all birds; hummingbirds are the only birds that can fly backwards. This is a group adapted for extreme flying.

Saturday 29th

I had done a wonderful thing. I had transported two soufflés from Norfolk to London, backpacked them along the Thames and then reheated them at my father's place. I was back after an excellent visit: and I found a marsh transformed. The sky was unbroken blue, the wind light, the sun warm. I sat there in shirtsleeves: a casual observer wouldn't have doubted my sanity for an instant. The headache was less insistent, my father was older but very cheerful behind his glass of Sauvignon blanc. Two blackcaps were singing to each other.

A falcon flew over, high. I always try to rise above the slight disappointment when a falcon hovers, or tilts in the air to reveal brick-red colours: this was one was a kestrel all right and a fine one too, never mind that he wasn't a hobby or a peregrine. Be grateful for what you've got: you never found Gerard Manley Hopkins complaining about kestrels. A cuckoo called, just once. I kept my eye on the hovering kes: how well could he see the ground from 200 feet?

I shortened the focus drastically to catch a white butterfly and it revealed itself as a female orange-tip (so without orange wingtips); at the same time I noted the contact call of a whitethroat. I felt mildly smug at this double feat of ID, though it's within anyone's scope.

An insect landed on my hand: a quarter of an inch long, the wing-surface lightly mottled in shades of browny-grey. It held one wing on top of another, but I couldn't tell if it had four wings or two. I could have obliterated its life with a flick of the finger, like one of King Lear's wanton boys, but refrained, Jain-like.

A sedge warbler was singing: my seat is too far from the heart of the marsh to hear them on a daily basis, and besides, they hadn't been going flat-out in the forbidding weather. Making up for lost time, he struck up again.

> 🍃 Kestrels can pick out a beetle at 200 feet. Like many birds they can see ultraviolet light: that enables them to see the trails of urine that field voles use to mark their favourite routes through tunnels of grass. The human sensory world is shared only by our nearest relatives.

Sunday 30th

It was bright, but so breezy I went back for another layer. All the same, this was May as it was meant to be: a chaffinch, a blackbird in song, a harrier in flight, the leaves on the ash almost fully formed: how do those tight, fat little buds produce those long, feathery fronds?

There were squeaky, hectic magpie sounds from the willow. This might be the very moment of fledging, but through the leaves I could make out only a lot of busy movement.

I wondered about the swifts and martins: and my wondering called them into being. A mixed flock of a dozen flew overhead: I followed one of the swifts through a series of zigs and zags: not a single flap.

A buzzard was doing its grand glide over the marsh, but there was a female harrier there too, and she folded herself into the W-shape to dive at the buzzard. The trees across the river were now so frothy they looked like an overflow of Guinness. Another swift crossed the marsh, this time with power full on, wings a blur.

And then I heard it. There had been hints and whispers before, but now, for the first time, the full song was loud and unmistakable and I rejoiced: because here was a willow warbler, not only a special favourite, but missing from the marsh for the past three years. There it was again, a sweet lisping descent down the scale, simple and perfect. It seemed at the same time impossible and gloriously inevitable. Consequently I rejoiced.

The martins were still overhead, calling to each other. Two swifts flew over, both on full power, their interest not in food but each other. A skyful of love, I thought sentimentally – and then noticed a male sparrowhawk climbing steeply, all agility and power. The martins vanished: the swifts were already miles away, rejoicing in each other.

> The term 'fledging' is a little ambiguous. In one understanding, it is the period between hatching and flight; in another, it is the moment at which the young bird leaves the nest with fully developed flight feathers. Either way, late spring and early summer is the fledging time, and once you have got your eye in, such events are quite easy to spot: the birds in a group, often calling, and with a slightly unfinished look to them; through binoculars you can often make out downy chick's feathers yet to be shed.

Monday 31st

Almost you might believe the summer would come. It was sunny and cloudless, a brisk breeze on my face. East? Northeast? The willow warbler was singing out with immense confidence and my headache was losing its grip. A return visit to the doctor would not be necessary.

In some ways the willow warbler made up for the loss of the Cetti, though I know that's a foolish way to think. When I began to learn birdsong one of the first songs I grasped was willow warbler and it was the entry to a new world: as the wardrobe was to Lucy, the willow warbler was to me.

A kestrel flew over the marsh and stopped to hover: the best weathercock there is, because they always hover beak to the wind. This one faced due east almost to the degree. A buzzard came gliding over and a crow leapt out as if from a trapdoor to take them both on, especially the buzzard.

Both moved on, but the buzzard came back a little later and went into an east-facing hover itself. It was a pretty good hover for a buzzard. The crow came back and chased it away. After another decent interval the buzzard was back once again; this time it was an oystercatcher that saw him off with lots of shrill piping. Hard work, being a buzzard.

> Willow warblers were once ubiquitous but have been in steep decline in England since the 1970s, though not in Scotland (you can get the figures from BTO). They are one of those ordinary birds that have become special. Hearing one was always a pleasure: now it's a special treat. I wish it wasn't.

JUNE

Tuesday 1st

The air was full of floating sallow seeds, the sky was cloudless, the breeze light. I seemed to have caught the siesta time of the year, the moment all species take a break, the better to appreciate the brief sumptuousness of things. A whitethroat sang a few bars.

I saw two thermalling buzzards; as they reached 200 feet

a crow noticed them as well. He made a steep flapping climb towards them, but this time he had overreached himself: in these circumstances the buzzards were at their best, and they had all the advantages of height. It was a brave ascent, but the buzzards had the upper hand or talon. His assault got a response of pure contempt and he folded himself into a D-shape to descend; as I followed him with the bins I lost him behind the leaves of the ash. The view before me was increasingly restricted by green growth.

I was working on a piece about the blue whale: tongue the weight of an elephant, body weight equal to 1,500 people, a suckling calf gains nine pounds in an hour ... so I thought of marvels nearer home: the wren not much bigger than a ping-pong ball that sings in the British winter, swifts and their three-year-long flight, caterpillars turning into butterflies ...

A bumblebee circled me, marvellous enough. I coughed: a white head rose from the dyke and one of the swans was glaring at me through the tangles of vegetation.

> You get these sleepy periods around the middle of the day in late spring and early summer, and they generally occur at the time we humans are most in the mood for a nice walk. You have a choice: get out good and early or enjoy whatever is up and about when you are.

Wednesday 2nd

The weather was still sunny and June-like. Two blackbirds were singing as I sat and then the willow warbler struck up. I hadn't heard him all the previous day and wondered if he had moved on, but there he was in all his sweet strength. There was a sudden outbreak of growling in the heronry.

Time brings depth and meaning to all things. I remembered the thrill of discovery when I first identified a willow warbler by song thirty-one years earlier; these days the pleasure is less sharp but infinitely richer. I have heard this song in Africa where the birds go for the winter; I have heard it in times of plenty; I have heard it in times of scarcity. Now, hearing one from my seat facing the marsh, I felt a pleasure I never knew when I first picked out the song from the joyous babble of the Suffolk lanes.

A chiffchaff struck up. I wondered which bird produces the most sound units in the course of a marshland spring, and which fills the most seconds with song. This chiffchaff was a serious contender for both.

But spring was settling down. The year was moving into a period of consolidation. The chiffchaff was no longer singing to establish a territory: he was keeping it safe while he and his partner used it for what it was claimed for: hatching eggs, feeding young, becoming ancestors.

A buzzard perched in the lone alder: the chosen twig bent like a boomerang but held firm. The crows left him alone for once. A wren sang close to my left ear and then the willow warbler sang out again.

> Discovering birdsong really was like entering another country: and you, dear reader, can enter it too. Simple exercise: next spring make a point of listening to blackbird and song thrush sounds online. Once you have identified them both for real, you will have your visa into the new country.

Thursday 3rd

It's hard to be contemplative to a deadline, but I was doing my best. I had agreed to do a filmed book-reading for a local arts festival, pleased to be asked and all that, but it put today's sit on a tight schedule. Was that very Zen or not at all Zen?

The air was still, thick, warm and packed with insects, hanging still, buzzing in my ear, a tiny one thinking about diving into my eye. A blackcap and a whitethroat were singing, the heronry grumbled and a cuckoo called twice.

After a slow start there had been cuckoos both loud and close: now the window was beginning to close, just a couple of weeks left. Spring was entering injury time: score now or go home a loser.

Perhaps there had already been some scoring. I hadn't heard a female, but that doesn't mean there hadn't been one around, making the briefest of trysts with an ever-calling male – if only to shut him up – before setting off in search of some hapless host.

A cuckoo was calling now, a little behind me. I heard another call dead ahead. Canst move i' the air so fast? Apparently not: two birds called more or less in unison, about a mile apart. The double-call seemed to put them both off: better to fly elsewhere and call than get side-tracked into a competition.

I picked out a marsh harrier through the fronds of the ash: my bleak midwinter post was now cosy and sheltered. The fat dandelion flowers had turned into clocks. Now there were seeds as well as flowers, newly fledged birds as well as song.

> Female cuckoos don't go cuckoo: instead they make a wild, rich bubbling sound. You can find it online, though it takes a little searching for: about fifteen notes going down the scale, often celebrating the laying of an egg.

Friday 4th

It's easy to make the connection between rain and life in June. I could feel the ground soaking it up as I sat snug in my waterproofs and I relished my traveller's privilege. Budget travellers have no option but to involve themselves with the culture they are travelling through, a truth I learned many times in many places. Now I was budget-travelling through the seasons, unable to escape the culture of climate.

These days the ash had leaves and was much better at holding water, releasing it in colossal drops when a gust of wind hit. The willow warbler was the voice of this good wet world: singing in the rain on the edge of his territory, in the willow to my left. One cuckoo called and the other responded while my rain-pelted hat provided the percussion.

Two blackbirds were performing a duet, each reminding the other of the boundaries they had established with the beauty of their voices. I followed a lone swift through the bins and that led me to a dozen more, all flying ridiculously fast as they picked out the insects that flew between the raindrops.

The landscape before me was filled with vegetation and insects: just what a willow warbler needs. There had been a real growth spurt and much of the place was now impassable; I remembered the hymn about soft refreshing rain.

> 🍃 The highest speed ever recorded for a bird moving under its own power in straight and level flight is for common or European swift: 69.3 mph.

Sunday 6th

It was warm and sunny; I could feel the humidity in between sharp gusts from the east. A great tit sang loudly as if it was March: probably on his second brood. My headache was now quite gone.

Once again I had picked a quiet and gentle moment. There was nothing of obvious excitement and so the small things carried greater weight than usual: a whitethroat sang, a green woodpecker laughed, three sharp calls from a crow, perhaps its sweetest love song.

A blackbird struck up. There is an implicit threat to other male blackbirds in this lovely melody, but no doubt the hen hears it as another love song. I could hear plenty of squeaky chattering from the big willow: young magpies finding their way in the world. A song of joy, no doubt.

An orange-tip butterfly came past, male this time, exceptionally lush. It was like seeing one for the first time: understanding him for the marvel that he was. I remembered brief, long-lost times of psychedelic drugs, when the most ordinary things in the world called up astonishment from the bottom of my soul, as Aldous Huxley, stoned on mescaline, rhapsodised about the folds in his trousers. It is always instructive to grasp the marvellous nature of the ordinary.

A shiny black beetle the size of a pinhead landed on my thumb: I removed him as gently as I could. I caught a flutter at my feet in peripheral vision: a moth that was a quadrant of pale grey with concentric stripes of darker grey: subtle and remarkably lovely.

> It was a common carpet moth, nothing a proper moth-er gets excited about. I made the ID by photographing it on my phone and checking the image out on an app called Google Lens: technology even I can work. This can be helpful with plant ID as well.

Monday 7th

It was a fine, sunny afternoon with a light breeze. Several extraordinary things were about to happen. First, my old friend Ralph was paying a visit. Can you imagine that? Talking to a friend in the same room? Having a guest in the actual house? Sharing food, sharing drinks, sharing laughter? This wild fantasy was about to become reality.

But that was nothing. After that Cindy and I and the boys were to travel to Bristol to see relations we hadn't seen for more than eighteen months: and two of whom we hadn't seen at all, for they had been born in the time of Covid.

And while all this was wonderful, I was reluctant to leave the marsh. Wanting two irreconcilable things at the same time is at the heart of the human condition.

A dunnock sang out with great clarity, and there was a small contribution from a blackcap. A woman in a flowered dress was standing tall and apparently floating through the air above the river at a steady seven knots. She was presumably standing on a boat that was out of sight below the reeds.

I could no longer see the water in the dyke in front of me, though I could hear a moorhen through the tangled vegetation. A buzzard flew high over the marsh. I wondered how much longer the cuckoos would be around; Cindy had heard a brief bubbling from a female, so at least one of the males had got lucky.

I watched an insect in the air before me, moving slowly about its business. I couldn't even name the family to which it belonged: the more you look, the more you must confront your ignorance.

A harrier called, breaking my reverie. I looked up sharply, realizing as I did so that quite 20 per cent of my winter allocation of sky had been confiscated by the ash. But I found the harrier all right, diving down in a tight W to harass a buzzard. Two crows came rushing in, anxious not to miss the fun; all four vanished behind the heronry. A group of three swallows crossed the marsh: still not too late to make a nest in our out-buildings.

> Ignorance was on my mind that day: I had just received a copy of *Illustrations of Norfolk Plant Galls* by Robert Maidstone: a labour of love if ever there was one. Plant galls are blemishes caused by parasites: oak-apples are the best-known example. You'll never know it all: which is rather the point.

Tuesday 8th

In the sixth form Ralph and I argued about books; he was for Lawrence and I was for Joyce. Today we paddled the navigable length of the local river and on to the pub, which turned out to be shut, a small example of Covid uncertainty. Undaunted, we paddled back. Ralph opted for a doze, I for a sit: feeling pleasantly proprietary about the view before me and the hidden waters we had travelled.

The marsh was almost solid with insects, so full of life it was dangerous to breathe. I thought of the Jains, who wear mesh coverings over their faces to avoid the accidental inhalation of insects: a life-affirming thought in the time of masks – and really, this bioabundance of insects was more joyful than all the headline acts you will read in these bulletins. 'But heaven and earth was teeming around them, and how should this cease?' A line from Ralph's boyhood favourite, *The Rainbow*: and I had to concede the brilliance of the grammar that makes heaven and earth all one.

There was a sudden splosh from the dyke: I craned forward and saw only concentric ripples, but a new (to me, anyway) hole in the bank suggested water vole.

Before me was a common blue damselfly, a male, a slim needle a little less than two inches long. I caught it in the bins: how can anything be so blue? A long-tailed tit was calling from the nearest willow, the magpies were shouting jollities from the next one along. All in all, a very decent bit of teeming: I hoped it would carry on doing so without me for the next few days.

> A proper field guide to insects is a daunting thing to a beginner: far too many species. I made my first adventures into this unfamiliar group with the help of a beginner's guide, patchy and selective but hugely educational: *The Michelin Guide to Insects*. I would award the common blue damselfly three stars. There's a series of Concise Guides offered by Bloomsbury and the Wildlife Trusts: each one a great introduction to any unfamiliar group.

Monday 14th

I had been gone for nearly a week and the place had changed in my absence: the grass was now higher than the arms of the chair and my view was even more restricted. But as I sat down the elegant willow stump was before me, so were the lone willow and the lone alder, the heronry stood to my left and the river lay beyond.

Tomorrow I would travel to Essex to address a group of carers and their carees for the excellent charity Stepping Out. I felt the usual stresses that accompany travel in these hard times and performance at any time: perhaps that helped to create the illusion that the marsh itself was shifting about beneath my gaze, as if I had been taking hallucinogenic drugs.

After a moment I had worked it out: a great air force of dragonflies was patrolling above the vegetation. It was like a vision of the future in a boy's comic of the 1920s: a sky full of tiny biplanes.

The nettles were higher, much higher; I could no longer

see the distant river embanking. Over the nearer nettles a red admiral was working assiduously; probably a female about to lay eggs. As I watched her a dragonfly turned and flew directly towards me, eyes like the goggles of Biggles and shining green. It was a Norfolk hawker, then, and as the name suggests a local speciality.

There were reassuring growls from the heronry, a female harrier dropped 200 feet in sharp dive; beyond her a dozen swifts. Summer was almost upon us.

> Nettles are an important food-plant, used by the caterpillars of red admiral, small tortoiseshell, painted lady and comma, all gorgeous butterflies. You don't get butterflies unless you have caterpillars: if you have a garden and want butterflies, cultivate a nettle-patch.

Tuesday 15th

I took on the charity event from duty and found pleasure as I talked to the carers and the cared-for about the joys and consolations of nature. A break in the claustrophobia of intense caring is as refreshing as rain in the desert: many of these people were caring for a stricken partner, sometimes chair-bound, sometimes with dementia. As we all know, those we love most are more capable than any other of driving us nuts: but here was a day out, people to talk to and the sun shining on us all as spring moved on towards summer. What's more, the trains were good and I was back in time to seek a little consolation myself.

At half-past five on a sunny afternoon there was a different feeling to mid-afternoon. It was all much busier: a blackbird sang very close, a wren, a chiffchaff, a robin – and then a moment of joy.

A male harrier – of course it was a harrier – turned in the air to catch the lowering sun and lit up as if he had thrown a switch: chestnut, sable, and argent. I heard myself making a weird grunt of delight as I sat back and a foolish smile crept up my face. I had been telling my audience how nature makes bad times less bad and good times still better: for the millionth time in my life I had proved the truth of this assertion.

It was, I suppose, a moment of enlightenment. Years ago I had attended a lecture from the Buddhist and High Court judge Christmas Humphreys. In the Q&A he was asked: 'But if you reach enlightenment, how do you know?' 'You'll know,' the old boy reassured. And maybe I knew, then. But if I did, I knew it wasn't a permanent state of being. So I kissed a joy as it flies: and was perhaps kissed back.

Above the marsh the cirrus clouds looked like a nineteenth-century engraving of a trilobite. A buzzard perched in the lone willow, unattacked. There was a second male harrier, very high, and then a female over the river: birds that had been extinct in this country were coming along in threes like the 49 bus.

> Marsh harriers were shot to extinction in this country during the nineteenth century, along with goshawk, honey buzzard and white-tailed eagle. Marsh harriers came back as the pressure eased, but pesticide pollution reduced them to a single pair by 1971. With changes in legislation and top-quality conservation work they are now back in decent numbers.

Wednesday 16th

My exposed viewpoint of six months back was now a cosy den roofed and walled with leaves. I sat in shirtsleeves rather than seven layers. I had no need of my snood. The idea of sitting here in a blizzard and living with seven hours of daylight instead of nineteen was unthinkable: the sun was shining above the leafy ash while blackbirds sang out over the impenetrable marsh.

An oystercatcher called sharply: it was attacking a female harrier, so presumably there was something worth protecting. If so, these birds have done a remarkable job: the price of oystercatcher chicks is eternal vigilance.

Two swallows flew across the marsh. A buzzard flew out of the lone willow but I felt no shame at missing it before it did so: it was too leafy to see inside. A little egret, freshly washed, flew easily from left to right; as it vanished over the alders the buzzard broke into a clumsy hover. There was another buzzard, more distant, and further still, tiny hints of movement in the air. Swifts, I bet.

> 🍃 Those of us who live in the seasonal lands tend to reserve our amazement for rainforest and desert, blinded by habit to our doorstep marvels. 'Most of our faculties lie dormant because they can rely upon Habit,' wrote Marcel Proust. This project is about trying to break Habit so that we can see deeply familiar things for the first time.

Thursday 17th

One of those drastic alterations that our climate is so noted for: heavy overnight rain had continued deep into the morning and the grass had bowed to its supremacy. As I walked to my spot my ankles snagged and my legs were splattered but I was back in full waterproofs. It was still overcast, but warm enough. A blackbird was singing; a second joined in; a chaffinch tried to drown them both.

I peered at a garland of wild pink roses that had risen above the tangles on the marsh, feeling the weight of my eyelids. Hard morning's work, sure, but if I fell asleep here I'd be overgrown within the hour and no one would ever find me.

Was that a sedge warbler behind the chiffchaff? No, it wasn't. For a while I tried to hear it as a grasshopper warbler, a classy bird that has yet to turn up here, but it was nothing like. After a few dozy minutes I pulled myself together: this was the grunting sound that chiffchaffs sometimes use to punctuate their chiffing and chaffing.

A whitethroat appeared without warning on the kingfisher perch, jaunty and confident. It flew to the willow stump and foraged there briefly before moving on. Behind the heronry a female harrier was working her way through the wet, heavy air.

> 🍃 Dead wood is an important food source for many insects, and these insects are an important food for many others; it's a good idea to keep an eye on such places.

Friday 18th

It was what we humans call a rotten day, but it had perked up the marsh no end. There had been a great deal more rain, and though it stopped by mid-afternoon as I stepped out, everything was dripping and two shades darker in the murky light, like your trousers after a walk through wet grass.

But I was waterproofed up and able to delight in a sharp increase in birdsound that came with this pause in the rain: whitethroat, wood pigeon, young magpies, a buzzard yowling and a chaffinch singing. Three swifts flew overhead, hawking and swerving; a lesser black-backed gull flew along the river.

The wind was blowing in sharp gusts that riffled the reed-bed as if it was a field of corn. A female harrier flew across, looking bedraggled, as if flying was hard work. No doubt it is when you have a nestful of chicks to feed. A dunnock and wren sang out simultaneously, the dunnock almost matching the wren for volume.

A male harrier turned up, looking relaxed and elegant, in sharp contrast to the female. Maybe I had just caught her at a bad moment: a bad feather day. A second female dropped in and perched low on a hawthorn. After a pause she flew downriver, climbed with a couple of easy flaps, turned and cruised back. The wind blew a few heavy drops from the ash and then the rain got serious again.

> Dunnock and wren are important songs to get clear in your mind: simple songs that provide a great deal of the backing track to spring and summer. It's a good idea to give yourself occasional revision courses on these two until you have them sorted; a birdsong app on your phone is the easiest way to do this.

Sunday 20th

The solstice would take place in sixteen hours: at 0431 the following morning the northern hemisphere would start to tip away from the sun, the daylight hours would start to reduce and summer would begin.

The rain had continued through the night and on into the morning again and the world was dripping: three-foot stems of grass were now shallow arches pearled with raindrops. My seat seemed to have sunk deeper into the earth.

And it was all very slow, for summer is a slow season: a hint of chiffchaff, dunnock, magpie. I sat without taking notes or thinking about words, just breathing wet air. If I craned my neck I could see an elder tree frothing with flower.

I followed a pigeon with the bins: was I reduced to making notes about pigeons? I could pick out dancing motes of dust which turned to reveal sickle-wings: distant swifts. A pigeon crossed the marsh in a characteristic series of switchbacking glides – yes I was – a display flight showing off the white wing bars and bright white neck patches. It made four glides before disappearing behind the heronry.

The heronry grunted and a cormorant took off from the river. Why do I always think they look like pterodactyls? It's not as if I'm familiar with pterodactyls. Something to do with the awkward stuck-out neck and the long, blunt beak, along with the gliding effectiveness of their flight.

> Summer is a season without arrivals and departures of birds, without obvious drama for obsessive birders. But there's plenty going on at a more subtle level, and other groups – especially butterflies – take centre stage. The world contains greater riches for the all-rounder: most birders chase butterflies and dragonflies in the high summer.

Monday 21st

Summer was eleven hours old. Once again there had been heavy overnight rain and it was still overcast. I was feeling slightly boggled: I had done three telephone interviews, all for my *Radio Times* sports column, and you have to put a lot of energy into such things. It was good to listen to the blackbird and the robin and the sound of the wind, blowing into my face from the northeast.

We associate summer with sunlit days, but it's more about daylight than absence of cloud. The long hours of light feed the plants and they allow everything else to live.

A lone swift was banking first one way and then the other, always at speed. Was it in touch with other swifts? Did it know where its friends and colleagues were? Perhaps this singleton was not alone after all, relating to distant birds I couldn't see.

A buzzard flew over me towards the river in a long glide. A second appeared, calling, and the two danced briefly around each other – a do-si-do – before one dropped into the lone alder and the other continued. The first bird then flew again, circling the marsh three times, gaining 30 feet without a flap.

A female harrier made the same crossing a little later, directly into the wind, but she had to flap: she needed half a dozen wingbeats to get across. Her lighter body made her less effective at gliding into a stiff breeze: a buzzard can weigh nearly 1.5 kilos, a female marsh harrier is half of that at best.

> People studying giraffes have a hard time working out whether a giraffe is associating with another giraffe a mile away, for in giraffe terms – up there you can see forever – a mile is close. Perhaps it's the same with swifts.

Tuesday 22nd

It was still overcast, but a brighter sky to the north promised better things. The wind was gusting to around 30 mph and its sound filled my ears, competing with the blackbird and the chiffchaff. These two were as dogged as ever, a song for every kind of weather.

A magpie landed on the kingfisher perch: its short tail marked it as a young one, doubtless fledged a couple of weeks back from the nest in the willow: how little time it takes to produce this slick, pied, confident-looking creature, already, it seemed, knowing every trick.

Over the river a lesser black-backed gull cruised comfortably, neither its temperament nor its feathers ruffled. Long-tailed tits were calling to my left: another fledge-out. The blackbird and the chiffchaff sang on.

A female harrier crossed the marsh in a single unflapping glide: but she was moving at 90 degrees to the wind and now her lack of weight was no handicap. She was very busy, using the wind to make a series of fast glides, savouring her skills. One flick of the wings and she was 100 yards further on.

> Flight is all about air passing over the wing-surface and creating a difference in pressure. You can demonstrate this for yourself by cutting a strip of paper a couple of inches wide and holding it to your lower lip. You then blow: and the paper rises. It flies. You have discovered flight.

Wednesday 23rd

Had one of my neighbour's big red cows crossed the dyke? That would be an excitement. A few years back I had looked up from the screen to see 100 sheep round my hut. But no, as I peered through the ever-loftier vegetation it was clear she was still on the wide, wet pasture that leads to the river. The sky was blue, streaked with cirrus, cool and pleasant.

A buzzard flew, its wings extended so far from its body it

seemed to be playing eagles. Two chiffchaffs were duetting; a whitethroat was getting up a head of steam. I could hear the young magpies finding out about the world: it seemed to please them.

A hidden moorhen called from the nearest dyke: not the usual sharp alarm call but a gentle squeaking: calling to an invisible flotilla of young. I picked up a movement in the sky, found a buzzard – exit pursued by a harrier. A day without a harrier is a poor thing: daily harriers should be a basic human right. Or at least daily access to nature.

A blackbird sang two or three slow, measured phrases. Then the whitethroat struck up again: singing fit to burst, and in his ecstasy of song he took off and crossed half the marsh, singing his loudest at every whirring flap. This was a bird in his pomp.

> It's called a song-flight. It seems a mad tactic, an invitation to every passing sparrowhawk to have a crack. But the volume, the extravagance and perhaps the recklessness impresses other birds of the same species, both males and females.

Thursday 24th

A female harrier rose from the middle of the marsh as I sat: a privilege because harriers are hardly ever that close, it's as if we humans carry a 50-yard harrier-exclusion zone with us wherever we go. This one was almost confiding.

The weather was quite close as well: warm, humid, overcast. A chiffchaff sang. I could hear a male harrier calling behind me, but I could see nothing through the willows: a small section of the great roll of William Morris wallpaper that decorates the Broads in summer.

Two swallows made a pass across the marsh. There should have been many more of them, visible almost all the time, and certainly at least one pair should have been nesting in the stables, but none at all this year. Nesting hirundines should be another of those basic human rights.

A butterfly was flapping hard under the lower leaves of the ash: probably a red admiral. The blackbird was hard at it, but a blackcap nearby was much less fulsome: just an occasional soundbite, for the year was getting old. A cuckoo called, the first I had heard for a few days. He called just twice more: distinctly half-hearted after his earlier excesses. Or perhaps just sated.

> The cuckoo was probably Grove, who had been satellite-tagged by the British Trust for Ornithology three days earlier just a couple of miles away. I planned to follow him to Africa in a few weeks.

Friday 25th

I really thought the chunky brown raptor was a female harrier. I had sat down on a still, overcast day after an unexpected glut of journalistic work. I heard a male harrier, turned and saw him with this second bird, which I put down in my notebook as a female harrier; they seemed to be having fun together as they flew by rapidly and vanished behind the heronry.

On the far side of the river there was a flock of wood pigeons: thirty of them. We were barely into summer and yet here was an irrefragable sign of autumn, if not winter. These birds were not (or perhaps no longer) interested in being half a pair: they wanted to be members of a flock, finding safety and food in numbers.

The blackbird was singing well, as always: this really was a rather special bird. A heron flew into the heronry, his crop bulging, confident of an enthusiastic welcome. Two red admirals flew combatively into the ash; a few moments later one of them descended, somewhat battered.

A buzzard emerged from behind the heronry – and right behind, making that oddly calm mobbing call, the male harrier. So it wasn't a female harrier at all, and this wasn't a loving interaction but a skirmish. Easily confused.

I made the correction in my notebook, looked skyward and found a hobby. Probably. It looked right and it didn't let me down by hovering. But – well, I could have been mistaken.

> Learning wildlife is about making more and more mistakes: ever better mistakes. Ignorance is an incentive, and misidentification is both an inspiration and an investment for the future.

Sunday 27th

Once again the song of blackbird. Perhaps this bird would never stop: certainly I would never want him to. A steady breeze came from the northeast with sudden fierce gusts. The magpies were calling all around; two of them appeared on the willow stump with that essential magpie air of being up to no good.

A second blackbird joined in from the right, fine for melody but lacking the panache of the main singer. Another flock of pigeons flew by; ten of them, as autumnal as a falling leaf. A lesser black-backed gull flew along the river and I watched it give a sudden all-over aerial shudder: it had been ducking into the river and was shaking off the excess from his oiled, water-tight feathers.

A buzzard was flying high, a good 500 feet, only just visible to the naked eye. Four crows flew across, intent on some mysterious but very noisy business of their own. Parents with young ones, I reckoned.

> Most bird species maintain their feathers with the help of a fatty secretion from their own bodies. This is important because feathers are essentially dead structures, made from keratin like your fingernails. Preen oil helps in waterproofing feathers and controlling parasites.

Monday 28th

I live in a wonderfully watery landscape but in terms of rainfall it's the driest place in Britain. Or at least it's supposed to be: heavy overnight rain had led to a morning of light drizzle, and now it was overcast and cool, my seat wet. But the blackbird was singing beautifully, a female harrier was climbing above the marsh in wide circles, I was wearing waterproof trousers and all was right with the world.

The second blackbird joined in again; maybe I had underestimated him, because this was a pretty good effort. Human rivals in pubs and political platforms exchange insults and ugliness, but the dispute before me was gentle and lovely: each bird forcing yet more beauty from the other.

Then a song thrush started up. It was as if we had shifted back to March. No doubt he was investigating the possibility of a second brood, but it felt like a rip in the fabric of time. The wood pigeon flocks said autumn, the song thrush said early spring, the calendar said summer.

Three crows flushed a buzzard from the lone alder; it's hard for a big raptor to find peace around here. A male harrier flew by, very close and unharried by crows. Two big umbellifers had grown up in front of me, standing proud of surrounding vegetation.

> Not omniscient, I looked up the umbellifers back at my hut; they were wild parsnips. Checking things out after a sit or a walk gives added meaning to the experience.

Tuesday 29th

The sky was silver, but only parts of it had been polished; the rest was rather tarnished. The better blackbird was singing through the light drizzle, this time from the weeping willow in the garden, and there was a chiffchaff in good voice.

The wind, still gusting in my face from the northeast, was making two noises. The first was the roar of the wind itself; the second was the air passing my ears: they were being played like a couple of flutes. If you're a mammal with external ears this double effect seriously reduces the information you receive through your senses and makes a prey animal more vulnerable, and it's why horses are spooky on windy days.

A goldfinch crossed the marsh, calling prettily, scattering those tinkling gold coins. The polished bits of sky had almost all vanished but the blackbird struck up again and the chiffchaff hadn't stopped. And what was that? It sounded like a reed bunting, but it couldn't be; not enough reeds.

Suddenly, and this time unmistakably, a Cetti called. Just once. I longed to think that this was the bird who had called so often earlier in the year, silently rearing untold numbers of Cetti-ettes, but common sense told me this was an unpaired male passing through.

Damn it, it *was* a reed bunting the second of the year from this spot. A swift fizzed over, visible for half a second, returned a moment later but this time in the form of a hobby: a metamorphosis brought about by my initial error.

> Habitats change over time, and the marsh was getting reedier. Reed warblers started to breed there four years back: perhaps reed buntings would do the same thing the following year.

Wednesday 30th

The chiffchaff was still going strong and there was an unexpected yelp from a little owl; it was cool, damp and breezy after overnight rain. Above me I heard a thin, high, steady whistle, repeated again and again: probably a treecreeper, birds with a special liking for the ash.

A female harrier was trying to use the breeze to hold still, but she couldn't really do it: up and down like a yo-yo. Eventually she flicked her tail and gained a rapid 100 feet, where she was joined by an oystercatcher. Irritated by its piping attentions, she drifted away.

Two whitethroats were singing competitively on either side, while a dunnock sang his only phrase as if he had just thought of it. I gazed at the wild parsnips: their shape – if not their size – reminded me of the emergent araucarias I had seen in Atlantic rainforest: trees in the same genus – *Araucaria* – as the monkey-puzzle tree. This was the pseudonym of the Reverend John Galbraith Graham, crossword compiler for *The Guardian*: I wished I could have passed this fact on to my mother, who loved a good crossword.

Two whitethroats came out to forage on the willow stump, very close to each other: a pair, presumably, rather than the singing rivals. A buzzard hung in the air, gliding at the exact speed of the wind, making a far better job of it than the harrier. I waited a little longer in case a kestrel came along to do the same thing better still, but he missed his cue.

> You will often hear such a thin, high whistle in woodland. It's an alarm call used by several species and it's very hard to locate: a shared signal designed to beat predators. But I'm pretty sure that one was a treecreeper.

JULY

Thursday 1st

Would I ever see the sun again? It was as overcast as ever but at least there had been no overnight rain. A young buzzard was hunger-calling repeatedly: a sound that expresses the bird's need for food and also soothes the parents: they know where the young bird is, and that it's safe.

The air above the lofty vegetation of the marsh was thick with insects, many of them mosquitoes. And was that a chaffinch sounding like a goldfinch or a goldfinch sounding like a chaffinch? You can spend an age puzzling such matters if you have a comfortable seat and no urgent appointment. A friend had described a bird in her garden as sounding like one of Joseph's guitar solos; I correctly diagnosed this as goldfinch and tried without success to get Joseph to write a goldfinch étude.

I felt – or did I? – mosquito feet on my face and flipped my fingers in non-lethal discouragement. Should I sit here and wait for one perfect bit of wildness, as I used to wait for the one perfect wave when body-surfing in my Asian days? Ah! It was a goldfinch: he had stopped playing chaffinches to concentrate on the real thing.

The previous day walking around the marsh I had heard willow warbler and Cetti. It had seemed for a while that both species would breed here for the first time in a while, but both had failed. This had saddened me, for personal and for universal reasons. Maybe next year. These two were just passing through, but at least they kept that possibility alive.

> There are thirty-four British species of mosquito; most of them are nectar-feeders in their adult form, though a few take blood from birds and mammals. Which includes us, of course.

Friday 2nd

I stepped out in my shirtsleeves, legs free of waterproof trousers, to sit on a dry seat because – wonder of wonders – the sun was shining. The chiffchaff sang as if rain and sun were all the same to him; a blackbird murmured appreciatively. I was buzzed about by flies, all of them responding to this thrilling new warmth. I felt a small itch in my calf: souvenir from a mosquito, no doubt. A red admiral flew overhead and a little owl called.

Another red admiral came bundling across, swift, direct and purposeful: red admirals do not have butterfly minds. A kestrel arrived a couple of days too late to show off his gliding skills: not enough wind for the stationary glide today.

There was nothing much going on – but how pleasant to be sitting in the shade of the ash, the warm sun on the marsh, a cloud of insects over the parsnips, the vegetation and the air filled with life. How pleasant not to be looking at a screen, to be looking at something far away, living and yet not human.

I remembered a throwaway line in Frans de Waal's superb *Peacemaking Among Primates*: researchers at the chimpanzee colony at Arnhem Zoo in the Netherlands, apparently studying the chimps, were actually observing the human visitors. They found that the most frequent remark was 'Oh, I could watch them for hours'. The average length of stay was 3.5 minutes.

> I would like to claim moral superiority over the bad chimp-watchers, but that would be wrong. The fact is that by consistently staying longer – by consistently going through the going-through phase – you create in yourself an ability, a desire, and even a need to stay longer. And we can all do that.

Sunday 4th

The previous day a melodramatic storm had come flashing and banging in, dumping the rain in bucket-loads. But today was like the last movement of the Pastoral Symphony with its 'thankful feelings after the storm': a mood shared by me and everything else in the pleasant late afternoon sun.

The blackbird – who else? – led the way, the air around us filled with the soft sounds of insects. The storm had inspired a drastic surge of growth and my view was even more limited. The parsnips were now at eye-level as I sat.

I picked up the movement of a sizeable bird through the spokes of nearer parsnip: female harrier. I stood up to see her properly – perhaps I should jack the chair up about three feet – and watched her arc across the marsh and disappear behind the lone willow. After a long moment she reappeared – on the same side she had gone in.

Blackbirds sing many clean, pure notes, which is why they are such beloved singers. But they mix them up with distortion and fuzz. Guitarists like Joseph will do this with the flick of a switch; human vocalists can make an adjustment to 'sing dirty'. When Freddie Mercury and the coloratura soprano Montserrat Caballé sang 'Barcelona', Freddie, singing pure notes at first, went majestically dirty for the big finish. The blackbird above me was doing the same thing.

> When a watched bird flies behind a tree it often fails to reappear on the far side, or even at all.
> An observer must adapt to avian rather than human logic.

Monday 5th

After a sunny morning it had clouded over and the wind was blowing urgently from the south. A female harrier got up and flew across the marsh; as she went past the lone willow she began to move laterally while still facing forward; I have often asked a horse to perform the same sort of manoeuvre, one that requires balance, control and athleticism (mostly from the horse). The harrier went behind the lone willow and didn't come out the other side. The sun broke through and warmed my back.

Then the harrier was back, dealing with the wind with her usual airy confidence, making a series of tiny adjustments for the gusty conditions. Another female harrier appeared from nowhere and attacked, talons out: for a moment the two grappled in the air and then separated and vanished in different directions.

The sun went back in and the temperature dropped like a brick. A male harrier flew the line of the river: perhaps partner to both females, father and provider to two sets of chicks.

The air above the marsh was buzzy and busy. A blackbird tried a few easy phrases with some dirty notes thrown in. Four linnets flew over, bouncing in the air – and then a faint line in the sky above the heronry and I knew it at once for a marsh harrier. My months of sitting may not have brought me to full Zen enlightenment, but they had certainly sharpened my eye for a harrier. I felt as if I could pick one from the sky by willpower alone.

> They may not have been two female harriers. Subsequent reading and thought made me wonder if the attacking bird was a newly fledged chick, play-learning about flight: preparing for a career in the sky.

Tuesday 6th

I was in an experimental mood so I tried to will a harrier into being. It was cloudy and cool after yet more rain. I sat there in my waterproofs but the sun came out to mock my precautions. The chiffchaff and the blackbird sang on, constants in a strange summer. No harrier. Damn.

A red admiral and a meadow brown whizzed past – and then of course I had my harrier, briefly – very briefly – in sight near the heronry. A magpie flew onto the willow stump and with careful, finicky movements wiped its beak a dozen times on the branch at its feet. Behind it a dragonfly, perhaps a Norfolk hawker.

The magpie looked in my direction, eyes alight with mad keenness, and flew off. The harrier – I knew I was right – emerged again from behind the heronry and flew across the marsh, rapidly gaining height with the occasional flap. She then went into an easy dive, vanished behind the lone willow and didn't come out again.

But the male harrier did, flying with more urgency than his presumed partner in the direction the female had come from. Harrier chicks don't feed themselves, even if they can fly a bit.

I looked up into the ash. How much ground would the leaves cover, if they could all be removed and spread out beneath the tree? These leaves that didn't even exist a few weeks ago; now they could hide several football pitches. A whitethroat flew onto the willow stump, looked around, flew to a higher perch, did the same thing, and repeated the whole process once more from the summit.

> There are various formulae for calculating leaf area, which are either too simple to be accurate or too boggling to a non-mathematician. The point, though, is the different design principles for animals and plants. Plants (mostly) need a large surface area for effective photosynthesis, while animals (mostly) need to economise on space. Animals are maximal inside; plants are maximal outside.

Wednesday 7th

I was determined to catch him. Catch him in the bins, I mean. A decent-sized dragonfly was patrolling the dyke, or rather patrolling his territory. I wanted to see him close, partly to see if he had green eyes and was therefore a Norfolk hawker, and partly for the sport.

It was breezy, but the sun had come out to greet me as I sat down. There were many dragonflies above the marsh, but I tried to concentrate on the closest of them, constantly advancing beyond the blackthorn on my left for about two yards before doing an about-turn. I stared hard at the spot he had just left, bins at the ready – but once again he was too quick, no doubt enjoying the sport himself, flying like a Sopwith Camel dodging the guns over the Western Front. The chiffchaff from the ash joined in with the chiffchaff of the willows.

Got him! For less than a second I had the dragonfly in focus and the eyes were a couple of emeralds. A goldfinch sang out, his song half-drowned by the rising wind, but the turbulent air didn't seem to bother the dragonfly in his fast stop–start patrols.

It was a quiet afternoon for all that, so much so that I made a note about a couple of wood pigeons. Then I made another about how nice it was just sitting here, despite the lack of action ... I seemed to be examining my responses in the manner of the two clever schoolboys in Julian Barnes' *Metroland*. After listening to Brahms' second piano concerto they noted: 'Exhilaration ... aspiring thorax. Confidence. Not smugness, though. More a sort of firm bonhomie.'

A small white butterfly passed the blackthorn and two stock doves flew over. I nodded at them, bonhomously.

> There are getting on for sixty dragonfly and damselfly species in the UK, and it's a slightly daunting group. Again, it's a good idea to start with something like *The Michelin Field Guide to Insects*, which offers a good selection of the commoner species. It's always about making a start. The British Dragonfly Society website is also very helpful.

Thursday 8th

I had a little time before going to London to visit my father, so I was out there beneath the ash once again, seeking calmness once again, listening to the chiffchaff and the blackbird. It was cool and cloudy, a very high gull crossing the gull-coloured sky. The blackbird seemed to be singing instructive verses about the futility of anxiety; I did my best to listen.

A big hoverfly was apparently perched on the air in front of me, and, not for the first time, I wished I was an expert on this enchanting group: or at least able to pick out one or two species. I know: the solution is in my own hands. The thought of acquiring knowledge is intoxicating: and the rule counts double for wildlife.

Nothing hovers like a hoverfly: so perfectly still it's as if the air were amber and the insect caught in it for all eternity. But they can move from stationary hover to rapid and unpredictable flight in an instant of time, a purposefully deceptive shift like a winger bursting into the penalty area.

I had a half-second glimpse of a female harrier, though here I must ask you to trust my newly acquired super-power of harrier-vision. A bumblebee went past at improbable speed, the wind in its tail – and then a falcon was flying briskly across the marsh, its talons full of food, full of death, full of life, and from its chunky, muscular shape it was clearly a peregrine.

I took that as a rather ambivalent omen and got up to go to London while the blackbird sang on. My father would like the peregrine: we would remember peregrines we had shared.

> 🍃 Hoverflies have only two wings (most insects have four). They're related to houseflies and are perhaps even better fliers. There are 6,000 species worldwide and about 250 in the UK. That uncanny hover is the best way to pick them out from the rest. Many are brilliant mimics, and we'll look at that in a moment.

Friday 9th

I got back from London – good visit – with some urgent work to do, so naturally my computer decided this was the moment for a leisurely update. Why do they always do that? I was forced outside on a cool day of high cloud; two magpies were exchanging views in the top of the ash and two harriers flew beyond the river.

At the same moment the blackbird and Cindy's power tools both struck up: two artists at work. If you are making something of no obvious practical purpose to please another, you are making art, are you not? Certainly both were making something beautiful.

I felt a single drop of rain on my hand. A wren sang his introductory phrase only, no concluding trill; a whitethroat sang for a moment and then went into call, hurried and whooshing. The time for singing was passing before my ears.

I focused close, the nearer parsnip sharp as a knife: an insect rose up, clearly a hoverfly from its flight, but in the bins I could see every hair on its furry body and could hardly believe it wasn't a bee. That's another reason why hoverflies hang in the air with, it seems, scant regard for danger: they have a deep trust in their ability to mimic more dangerous species.

A pamphlet saved from *BBC Wildlife* magazine years back allowed me to narrow this one down to maybe three or four possibles: call that an advance. Above the dyke a dragonfly appeared, and was at once chased off by another.

> When a harmless species imitates a dangerous one, it's called Batesian mimicry. Many hoverflies look like wasps and bees, but their flight is quite different.

Saturday 11th

Sorry to bring this up, but I was thoroughly out of sorts. I know, I know – a guest should never bore a host by grumbling and a writer is always a guest in the reader's world. But here I am, trying to write about the healing power of nature, so I must present myself occasionally as a person in need of healing. When in Bristol I had practically trepanned myself when climbing an unfamiliar staircase, and I had been suffering from headaches ever since. My weekly plan is to spend Saturdays with Eddie, but he generously understood my need for a break.

So I sat out on a day so humid you could almost drink it. The chiffchaff welcomed me back, a loud wren sang out and a little owl yelped ... and then came precisely the sort of generous moment I had been looking for: half a dozen swallows chattering excitedly as they made their way across the marsh in their eternal curves, skywriting C and O and S. They weren't feeding:

this was fun, a recent fledge-out of young birds trying out their wings and finding them pretty damn good.

The wren sang out again: introductory phrase, concluding trill – and then a third phrase beyond that, proving himself a superwren, an *uberzaukonig*. Was this the reticent wren of the previous day? He was in full flow now.

I checked a possible hoverfly but it took off and flew like a bee. Here's a strange fact: my eye was getting better as my curiosity increased. And here's another: my curiosity increased as I sat stiller and for longer. I was no longer hurrying past most insects with a bonhomous nod at best: I was looking, noticing, seeing, contemplating, learning – and all without being aware I was doing so until that hoverfly turned into a bee and made wonder. Two of the swallows came racing back: for an instant they came together in the air and then parted. Why? Because they could. You know, I was feeling slightly less rotten about feeling rotten.

> There's a lot of stuff written about the many good things that nature will do for you. But I always like to remember that Alice found the Red Queen by walking in the exact opposite direction.

Sunday 12th

More or less with the contact of bum on seat I felt earthed: in the way that you make safe an otherwise dangerous electrical device. I could feel the bad stuff running out of my body, into the chair, down my legs, through my boots and into the long-suffering earth beneath.

The day was as overcast as I was, but at least one of us was brightening up a bit. The process was helped by a female harrier, always the bird I most want to see. Is it right to have favourites? But we are none of us impersonal observers.

A buzzard made a sudden low glide across the marsh: low is the preferred level of the harriers but this was a burly, fantailed buzzard all right. This slide down the air was leisurely, but there was a killing certainty about it, and when the buzzard failed to emerge from cover I knew there had been a moment of success: alas poor bunny. It had been a while since I had last looked closely at a buzzard: and as it rose again a little later, perhaps sated, I had time to appreciate this one.

Then a brief yapping and a male harrier arrived in a blaze of handsomeness, seeing off the buzzard with a bravura bit of flying. A goldfinch sang and two butterflies passed in quick succession, a small white and another with a flick of orange, probably a comma. Now for a full-stop.

> Commas are, yes, favourite butterflies of mine: bright coloured and with a ragged trailing edge to their wings. Their underwings are charcoal grey: with wings closed they look wonderfully like dead leaves.

Tuesday 14th

The previous day Matt Shardlow, chief exec of the invertebrate conservation charity Buglife, paid a visit with his family, and he introduced me to the marmalade hoverfly, which he assured me was the easiest hoverfly to pick out.

I was trying to do exactly that as I sat out on a sunny, breezy afternoon. A small white butterfly moved rapidly through the blackthorn; a long-tailed tit called from the fallen willow. A marsh harrier called; I looked up, expecting a male and found a female, usually much less gabby. A young buzzard was calling peevishly.

The harrier came back briefly: it's remarkable how much time harriers and buzzards spent crossing and recrossing; for the exercise, to maintain the boundaries of their territories, to hunt: in a quiet way they are almost always on duty. As if to prove my point the harrier reappeared once again, this time from the far side of the alders. She crossed the marsh, performed a sweetly economical 180 and headed back. I wished I was Gerard Manley Hopkins, about to write the best ever bird poem, not about a kestrel but a harrier.

Two chiffchaffs called, one on either side, as if I was their boundary-marker. I heard a sudden yell from a harrier: the buzzard was getting hell again.

> 🌿 The Hopkins poem is of course 'The Windhover', which is (among many other things) the greatest celebration of flight ever written, already sneakily quoted without attribution in these pages when I've been talking about kestrels. Science and art are not opposed disciplines; when it comes to nature they are usually on pretty good terms.

Wednesday 15th

There are moments when life changes: when you know that from now on, everything will be all right. They are there to be savoured even when you're too old to believe they will last. Cranial osteopathy on my poor head had parted the clouds: I felt as if I could dance in the manner of Blake's *Glad Day*.

But I was forced to move like someone doing tai chi, feeling that too sudden a movement would make my head fall off. I sat listening to the chiffchaff: only two notes and never for an instant tired with either. It's a song in monochrome: a limited palate, but so are Leonardo's red notebooks.

How does that bee hold onto the swaying umbellifer? It looked like the wildest fairground ride. Through the bins I could see that it was sharing the flower with a soldier beetle and two hoverflies. There was a burst of sun: perhaps my birthday would be sunny. I felt as if I deserved such a tribute.

A movement to my left: I turned my head sharply and was mildly surprised it stayed on. A red admiral was feeding on the brambles and a fat bee foraged alongside, both relaxed in each other's company. A small bird shot up 30 feet as if fired from a catapult and dropped straight back down; had to be a whitethroat, I thought, with all the certainty given to me by my hours of sitting.

I got up to leave and checked the nearer parsnip: two hoverflies were supping on nectar, unalarmed by my approach. After a moment I realized they were marmalade hoverflies.

> I am not showing off by naming that umbellifer: there's a margin of error of nearly 4,000.
> Umbellifers are a big family of flowering plants: their flowers are arranged like the spokes of an umbrella. The family includes parsley, carrot, parsnip, celery, coriander, giant hogweed and hemlock. A phone app called PlantNet will help with plant ID.

Thursday 16th

It was overcast and windy, the summer as elusive as ever, but the chiffchaff sang on, with a whisper of blackbird and some very nigh notes from a long-tailed tit. A swift, the first I'd seen for days, crossed in a straight line, going from A to B rather than feeding in curves.

A red admiral ascended into the ash; I realized that the blackthorn had grown four feet in the last couple of months, blocking my view still more. Soon I'd be looking through a keyhole. The patrolling Norfolk hawker went in and out of view.

A male harrier rose up on the wind and performed a series of dives and recoveries: display flight, shame to waste a good wind. Was this exuberance a celebration of a successful season coming towards its end? Certainty it was a reminder of who's boss around here.

The wren, the *uberzaukonig*, was back in his song of overplus: he too was not a bird to mess with. A buzzard came cruising along the line of the river and the harrier was up again to harry it off with three sharp passes. After that he made a decisive 180 with a flourish like a matador's cape: I felt like shouting *olé*.

> Swifts are only with us for a short time, arriving in late April and gone by early August: by this time of year every sighting is something to cherish.

Friday 17th

The sky was unbroken blue, red admirals and meadow browns were hard at it in front of me and a marsh harrier was calling out of sight. A whitethroat sang up. The usual word 'scratchy' helps with the ID but masks the musicality. Sure, he has hit the distort button, but that never hides a guitarist's virtuosity. I listened with the proper attention a serious musician deserves: as I did so I watched a harrier making a perfect dihedral and had a sudden pang that came, quite literally, out of the blue: yes, Zambia, the Luangwa Valley and its bateleur eagles, equally keen on the V-shaped glide.

At this point I took a break and paddled the river, coming back to my seat in good heart. The whitethroat felt just as good, giving it everything, flying across the marsh in full song to perch on the willow stump and sing some more from there. I remembered a performance of Monteverdi's Vespers I had attended with my father years ago – was it in Westminster Abbey? – in which a soloist had moved from platform to pulpit for a touch of added drama: *Laudate pueri*, praise ye the lord, oh ye servants of the lord ...

I needed nothing marvellous, for at this moment the marvellous and the humdrum could no longer be separated.

Even so, another marvel: a male sparrowhawk rose from the deep vegetation and crossed the marsh in his most thrilling way. After he had gone the blackbird struck up again, sounding a little weary, a little *fin du saison*.

🍃 Some analysts have heard subtle and complex mimicry of the songs of other species in whitethroat song. *BWP* says they also have a repertoire of twenty-five different calls. This is a very vocal bird, most often heard from clumps of bramble. And if you see a small bird rise above such rough vegetation and sing for about ten seconds, chances are it's a whitethroat.

Sunday 18th

It was later in the day than usual and my seat was in direct sunlight. A whitethroat was quietly waiting for me on the willow stump. Once I had made myself comfortable he went off in a song-flight. A deer called from a distance; a meadow brown went through the nettles in front of me and I caught a glint of orange too fast for ID: probably comma or small tortoiseshell. The blackbird gave out a phrase or two: subtle stuff. How much more did he have to give us? Every verse has been a delight.

The whitethroat was back on his stump and singing more emphatically. To my left another struck up with a series of longer and louder phrases; they seemed more complex as well. To my human ear he was the better singer and therefore a stronger, more experienced bird. But there was no action between them save these songs: the lines of demarcation had already been fixed, perhaps in favour of the better singer. Both knew where they stood or perched.

A buzzard came to perch in the lone willow and instantly both whitethroats shut up. A buzzard is not built to catch small birds but you can't be too careful with birds of prey. That's true even if you're a bird of prey yourself: the male harrier came piling out of the sky with whoops of glee. Exit pursued by a harrier.

> When you start to look at butterflies you will be shocked at how fast they move. We think of them as dainty, dilatory creatures: run after a hurrying butterfly and you will always come second.

Monday 19th

I seemed to have sunk deep into the earth, looking up at an ever-smaller patch of sky as it was crisscrossed by dragonflies. They looked enormous from this perspective: one bent his body in flight almost in half; another flew all around my hat, almost threateningly. The good whitethroat was singing to my left; I could hear the murmur of wood pigeons and the desperate barking of the dogs in their cage up the valley. The Bauhaus.

A heron and a harrier crossed the sky, fine big birds both, and I thought of the dragonflies of the Carboniferous era with their two-foot wingspans: a size impossible today. A goldfinch was fizzing and buzzing above me.

Tell me, dear reader: are you getting fed up with these dogfights between harriers and buzzards? Not as fed up as the buzzards. This one decided enough was enough: going talon-to-talon with the harrier a dozen times. Each swoop of the harrier was dramatic enough, but more impressive still was the recovery, regaining the height and the advantage in an instant, as if he was attached to the sky by a piece of elastic. It was like watching a fencer recover from a lunge back to the en garde position so brilliantly that the opponent can never take advantage of this moment of commitment.

Both birds were content to leave it there: further escalation would lead to serious damage for at least one of them. The buzzard backed down and retreated: the harrier's call of triumph sounded like mocking laughter.

> These extinct monster insects, precursors of modern dragonflies, are wittily called griffinflies: the record wingspan is 28 inches. They flew in air that was 30 per cent oxygen, compared to 21 per cent today: one reason why we don't have giant insects outside science fiction. The arrival of birds with well-developed flight in the late Cretaceous era made small size in insects an advantage.

Tuesday 20th

It was a lovely day, especially for the buzzard, for once having the marsh to itself, gliding across unmolested. A small white butterfly and a Norfolk hawker crossed in the air before: very different ideas about flight. The air was full of the sound of birds, but without the urgency of a few weeks back: a little bit of dunnock, a touch of stock dove, a good burst of chiffchaff, wood pigeon.

A butterfly flew from shadow into light, revealing itself as another small white. I noticed a stutter in its flight: a sort of hesitation, almost a trip. Another dragonfly flew past, made a

U-turn and flew back. They can do this 180 at full speed: it's like a Cruyff Turn, a famous trick in football that bamboozles a pursuer.

Butterflies and dragonflies spend a great deal of their short adult lives on the wing: if they weren't very good at flying they wouldn't survive. I have often tried to catch both in a long-handled net when they get trapped under the translucent roof of the veranda: they are equally hard to trap and release, but in very different ways.

The dragonfly's U-turn is obvious good flying. In contrast the butterfly trip looks incompetent, but it's as skilful as a comedian's pratfall. A butterfly's wavering flight is a highly evolved strategy for avoiding predators. That mock-hesitation is a big part of it: if you can't predict a flightpath, you can't intercept it. The buzzard, too often intercepted, made another pass over the marsh, savouring its solitude.

> Butterflies can be roughly divided into four main flying types: darters, flappers, jumpers and gliders: a really good observer (not me, alas) can identify butterflies from the way they fly. But with watching comes knowledge: I am better than I used to be.

Wednesday 21st

It was my birthday the following day, and we were to spend a couple of nights away in more northerly parts of the Broads. But I had time to sit out on the morning before departure, to see if I was to get a white-tailed eagle as a present.

It was bright and not quite sunny, high clouds diffusing the light. A blackcap sang a few gentle phrases and the chiffchaff sounded as if he would sing forever. My eyes shifted to a crowd of tiny black bees on the cow parsley. There's a strange scaling-down that operates as summer advances: birds are less visible while flying insects demand more attention. It's a bit like being Alice as she constantly journeys through size, but I can never decide whether I am becoming a giant, towering above the creatures I am looking at, or whether I am shrinking down to their size. Winter is about immensities: vast skies crossed by flying swans, while summer is about restricted views and tiny specks of life.

But then I heard a song I didn't recognize: and that's not a daily event, certainly not here. It was sweet and complex, with clear notes punctuated by trills. Surely a finch, yes, but what finch? Half a dozen swallows flew over the marsh followed by a few more.

Got it! Linnet. The first time I had heard one singing at our place: probably too wet for breeding linnets. They were popular cagebirds for the beauty of their song: and it was a treat to listen to this one. A Norfolk hawker flew straight at me, veering at the last moment.

> Birds – all wildlife – are about place: you're unlikely to find a cormorant in an oakwood. That's helpful: good to know what to expect. But birds – all species that fly – can turn up in unexpected places. I have seen woodland birds like redstarts on the seashore in the migration seasons: they have dropped down exhausted on the first bit of land they see.

Friday 23rd

A list of birthday presents: two cranes, two bitterns and an otter. The first two we had seen from a hired motorboat, the cranes in sight for a good ten minutes, the bitterns both on the edge of the reeds, and as I sat out with a last whisky, a just-visible otter came swimming through the blackness of the River Thurne. Now I was back with a more familiar landscape before me: its intimacy as thrilling as the exoticism of all other landscapes.

I could hear a young buzzard calling from the right: as I was trying to work out if the buzzard that had been calling from the left had moved, it started calling as well: an antiphon of hunger. Two soldier beetles were climbing a stem of grass, each intent on the other with the frankness that only a soldier beetle can manage. They are called soldier beetles for their red jackets but have acquired a modern nickname of hogweed bonking beetle.

The nearer buzzard was calling so piteously I wanted to go and feed it myself. From higher up the valley I could hear sheep, no doubt newly turned out. Sheep often sound like a very bad imitation of sheep: baaaa!

I could see more soldier beetles hanging in the air above the vegetation: unlike butterflies and dragonflies they are part-time aeronauts, clumsy but viable, forever searching for the great double bed of hogweed. It was good to be away. It was good to be back.

> 🍃 I have seen cranes at our place just once, an occasion when four of them flew over and circled for a bit, thinking about dropping in. Should you see one (try Hickling Broad or the Somerset Levels) tell them they're always welcome on this bit of marsh.

Sunday 25th

These summer storms can be very capricious. A Genesis deluge down the road was no more than a serious shower here. I went out a little later, the day richer for an early morning run on the kayak and a hefty session on the blue whale piece. I sat down, heavy-lidded and unambitious, the faint sound of goldfinch.

The buzzard was still hunger-calling to my right. I thought for a second I had heard a lilac-breasted roller – that's an African species, so perhaps this was the Luangwa Dreaming season – but it was just an unusual grunt from a pheasant. At least it made me concentrate: the chiffchaff started up again and a probable Norfolk hawker was patrolling the dyke. I made a note to that effect and then all at once the pencilled question mark shamed me. I was here to look, was I not? So what else should I do? So I looked and I focused and I waited and I looked again and eventually I had him plain: and it was the green-eyed monster all right. The pleasure I took in scratching out that question mark had been nine months in the making: nine months of sitting.

But this was mostly a day of windy silences and a sense of vague unease: not me, this time, but the sky. A red admiral checked my trouser-cuffs and found they were not nettles, and so unsuitable for the laying of eggs. Another storm was threatening: the marsh was battening down the hatches. I broke my rules and checked the Met Office on my phone: they were promising a belter.

Well, I was willing to sit through it. The world seemed to have paused forever, teetering on the edge of a storm. I heard a party of swallows charging through: I twisted my head to see a dozen of them between the top of the ash and the blackening sky.

> Weather affects wildlife. Birders who hunt rarities live with the weather like aeroplane pilots: October wind-shifts will have them hurrying to the special places where you never know what will turn up.

Monday 26th

Yesterday's storm missed its trajectory. Today it was cloudy and cool, the young buzzard calling hard to my left, as it had been doing all day. Cindy, hard at work with her power tools for most of the day, said if he'd been its mother, she'd have said all right, *have* some bloody cake.

There were soldier beetles hanging above the umbellifers. I got up for a closer look and found with them a small, wasp-like hoverfly: after a moment I realised that it was a marmalade hoverfly. I was becoming a world expert. The left-hand buzzard shut up, the right-hand one struck up and so the left-hand one had to join in. The one on the left called with a single note, the one on the right with two notes, perhaps a tone apart, sliding from one to the other in a glissando. (Joseph verified this.)

Six greylag geese flew by in close formation: as they reached the grazing meadow they disbanded with fine synchronicity, each diving to the grounds separately, spreading out as they did so: a Red Arrows manoeuvre called the bomb-burst.

A brief series of contact calls from the blackbird: not a hint of song today. It sounded like a unilateral declaration of autumn: an acceptance that the time of breeding was over and done. Successfully, I hoped: the world could do with more singers like him.

> The hoverfly's classic wasp disguise doesn't fool every predator. Bee-eaters catch wasps and remove the stings before eating them; when they catch hoverflies they eat them straightaway.

Tuesday 27th

There had been a thundery shower earlier, but all was still as I sat: cool, cloudy and, apart from a rattle of jackdaws, quiet. The soldier beetles were flying as if standing to attention, like James Bond in his jet-suit. Wood pigeons were calling to each other across the marsh and a moorhen squawked from the dyke: it seemed that things were going well beneath the tangles. There was a new pink flower there: later research revealed it as a great willowherb, or great hairy willowherb, a wet ground specialist.

Then, taking me by surprise, a song thrush struck up, but

not a song thrush as we know it. There was a lot of experimental stuff in this song, rough noises that sounded mechanical, some thrillingly pure notes and every now and then a perfect wolf whistle: all sounds I hadn't heard earlier in the year when the song thrushes were at their peak.

I wondered if this was a young bird, unpaired this year and now finding his voice. But as the song got louder and longer, I thought that surely no mature male song thrush would tolerate such a performance from an upstart. Perhaps it was an experienced bird trying out some new ideas: he'd been mainstream all season but now, with less at stake, he was relishing an opportunity to go rogue. Perhaps it was an investment for the future: verses that might play well next spring.

I sat on, wondering if this was intelligent speculation or flagrant anthropomorphism. A young marsh harrier was calling. There was a rumble of thunder, thoughtful and without serious threat ... now that really is anthropomorphism. From the Flood a burst of honking: the greylag geese were gathering. This was a sound of autumn: in the stillness of summer the year keeps on turning.

> Speculation is one of the wildlifer's pleasures, legitimate if you're not aiming for a peer-reviewed journal. Song thrushes are repertoire singers: the bigger their repertoire the more attractive they are to females and the more intimidating they are to fellow males. The best singers are the best survivors.

Wednesday 28th

A sit can't be entirely bottomless when you have a train to catch, but I had learned that every day is better for sitting – and so I was out there at eleven in the morning, trying to achieve enlightenment before the taxi came.

It was sunny and cool with a wall of clouds ahead. A bird flew over the marsh: without raising my bins or even my eyes I accepted it as a wood pigeon. Bad practice, I told myself severely, so I looked hard at the next bird. It was a wood pigeon. The geese were conversing loudly on the Flood.

A meadow brown hurried past, rather close: a butterfly has no time to lose, just a week or two of adult life: a very brief opportunity to become an ancestor, fulfilling the long time of waiting as an egg and a pupa, and between these stages, the long time of eating, as a caterpillar.

I heard human voices and remembered that Cindy had a client coming to buy some art. I was safely hidden: wouldn't do to look eccentric. There was the smallest patch of blue left in the sky: just enough to make a pair of sailor's trousers, so long as he didn't mind them tight. A green woodpecker called, an odd monosyllable and then, a moment later, four or five notes, still not quite right. This was a young bird, learning how to be a woodpecker. I sent him a message of good luck and stood up. I had a train to catch.

> Birds don't leave the nest as perfectly honed survival machines: they are living individuals, each one with much to learn.

Thursday 29th

Well, that was a fine old family time in London: a good gathering, both sisters, drinks, curry, laughter, my father in the best of form. It was a triumph narrowly achieved: my father, frailer than ever, had a last-minute crisis of doubt and decided not to go to the restaurant. My younger sister talked him out of this decision and the evening was a triumph: as merry a gathering as we have shared for all but two years. The following day I was in great good cheer as I took my seat though a storm of red admirals and meadow browns. It was sunny, the wind gusting up to 30 mph, hints of unsettled weather ahead.

My views were getting narrower every day: sitting here was like growing old. A female mallard fizzed across: how fast they travel when they have a mind to. A small white and a red admiral passed close enough to catch, had I been a bee-eater. This was a real butterfly day: proper warmth in the sun and the movement of butterfly wings everywhere I looked.

I watched the patrolling dragonflies wondering if they were the same individuals as before. A dragonfly can only hope to live for six weeks or so in adult form. There was a strange clucking and squealing from the ash: a bird I had never heard before.

More squeaks from the dyke: the moorhen family, getting bigger. I resisted the temptation to get up and look for them: after all, I was modelling myself on Sir Purun Dass in *The Second Jungle Book*, who sat so still for so long that the stag and even the great bear came to trust him. And then I solved the mystery: the strange sounds from the ash were just branches rubbing together in the wind. Sometimes the path of enlightenment leads to embarrassment.

> The more you listen to birds, the more you are aware of unfamiliar sounds: an unrecognized murmur can seem like a shout.

Friday 30th

It was eleven in the morning, just time for a sit before my old friend Lucy arrived to canoe the Upper Waveney with me. It felt like the wildest adventure: not the paddling but doing it in company. There had been rain, and more was promised, but I sat undaunted in the uncertain light. A small white blinked at me: for an instant the brightest thing in the world.

From the dyke I could hear parental calls of adult moorhens and excited squeaks from their unseen young: little balls of fluff that would be sitting on the water like bathtub ducks, the sleek adults close, for they are attentive parents.

A stock dove called, a hint of great tit. A female harrier turned into the wind and slowed, turned again and accelerated away: precise, economical and without ostentation, just as Lucy and I would drive our canoe. Perhaps.

Not much of July left. What would I do when the year was done? Would I still be sitting still here? Such a din there was from the dyke.

> There are walks, sits, paddles and other adventures that pass without major excitement: at least, nothing that sounds enviable when you tell it. But there is a pleasure just in being where wild things are: after a while you realize that there is no such thing as a dull day in a wild place.

AUGUST

Sunday 1st

This is the twelfth month I have named in these pages, but I was still a fair way from completing the circle – and as I observed the date I realized I was in no hurry to do so. It was late afternoon: the Upper Waveney had been navigated, Lucy's family arrived the following day, we had feasted and sung and all was right with the world.

The soundscape of the marsh before me was quite different from a few weeks back: the birds had stopped singing and instead we had the hunger-calls of buzzards, geese honking from the Flood, sound of herring gulls and the more clonky calls of lesser black-backed gulls. Spring's lease hath all too short a date, and now summer was cantering towards autumn. So it goes.

A wood pigeon struck up only to stop abruptly in mid-call, as it if had been hooked off the musical hall stage: that's not what the ladies and gentlemen have come to hear, kindly leave the tree.

It had been a big butterfly day, the buddleia in the garden alive with red admirals, the marsh full of gatekeepers and small skippers. Now as evening approached it was cool and quiet, butterflies waiting for a new morning. A minute of silence and then a female mallard yelled out, absolutely filling the place with her din.

> Nectaring insects make a beeline for buddleias: an Asian species that's gone native, one with straggly stems and great purple spikes of flowers. They provide a great opportunity to learn common butterflies on a sunny summer's day.

Monday 2nd

Make love not war, we used to say in the Sixties: but sometimes it's hard to tell the difference. Two red admirals flew past all wrapped up in each other, either fighting or loving. I could hear the sounds of carpentry from my neighbour's.

Over the dyke three dragonflies were involved in what looked like a three-cornered fight, perhaps a love triangle: it was impossible to tell as they vanished behind the blackthorn.

A peacock butterfly crossed the dyke, young and apparently freshly painted. Earlier that day I had seen three painted ladies on the buddleia: butterflies that travel from North Africa to be with us. I wondered if this would be a big year for them: there's something about this hard-travelling boom-and-bust butterfly that always lifts the heart.

A loose flock of 100 starlings fizzed across the sky and was gone in an instant: flocking birds, autumn birds. But a couple of swallows came after them, summer birds if ever a bird was: as I looked I found more, maybe two dozen at about 300 feet, busy specks of speeding, handbrake-turning life. But even they would be thinking about autumn and Africa and I knew how they felt. In the willow to my left a woodpecker started working with a steady *thwock-thwock*: more woodwork. In Trinidad they call them carpenter birds.

> The generation of peacock butterflies we see in the spring have over-wintered as adults; you sometimes find them in sheds and even cupboards, apparently dead but in fact biding their time. This fresh one would have been the progeny of an overwintering adult.

Tuesday 3rd

It was getting on for four as I took my place, pleasantly weary from a paddle, this time alone and on the local river. I could feel the sun warm on my left shoulder and the calling of wood pigeons sounded like an invitation to slumber. Nothing else seemed to be stirring.

From the Flood the herring gulls were in good voice: the flock display calls you hear at the beginning of *Desert Island Discs*. A little owl called sharply, perhaps a young one.

I watched a wood pigeon, there being nothing else to watch: it glided up to 100 feet and parachuted down, wings high, only ten degrees below the vertical. This gliding descent is the one manoeuvre wood pigeons can do with style, and they never tire of it.

A buzzard began hunger-calling: earlier in the day I had seen a buzzard, perhaps the same one, making these calls in flight. A green woodpecker called several times in quick succession; sounded like another youngster.

It was all pleasantly quiet, but just as I decided to leave I caught a hint of movement in the sky and found a female sparrowhawk crossing at 200 feet. Sparrowhawks always look cool: if they were humans they'd wear shades. Ray-Ban Aviators, no doubt.

> This pigeon rise-and-fall flight is usually for display, and when used for serious purposes it is accompanied by a couple of good, loud wing-claps.

Wednesday 4th

A large and startlingly blue dragonfly passed me on my way to my seat: too quick for detail and ID, or rather I was too slow. It was cool, sunny and breezy, and the place seemed full of a good quietness, a sense of achievement. That was the birder in me of course: that feeling of sitting at the still point of the turning year. For the dragonfly the season was just beginning and all kinds of excitements lay ahead.

But with the quietness of the birds there was a feeling that something had been accomplished: that it was time to pause, if only for a day, between the emergencies of spring and the emergencies of autumn.

As if to confirm that view, a mat of fluffy seeds flew by: the tree that bore them had done its job for the year. A kestrel flew over, pale and raggedy with one tail feather missing. It perched on the willow stump for about a second, changed its mind and flew on. This was all rather uncertain, rather un-kes-like: I diagnosed another young bird.

Then with great suddenness a flock of 300 corvids rose from the distant rim of the valley, above the oaks on the skyline: rooks and jackdaws in a great loose cloud, separating into two clouds and then joining up again to sink back down below the horizon. Here was the beginning of the great winter roost.

> The fluffy seeds were willows, which use the wind to spread them: a tree produces a fantastic number of them in the course of a lifetime. Seems wasteful, but if one of them grows up to become a mature seed-bearing tree, the tree has fulfilled its destiny.

Thursday 5th

I was off to London, happy for the chance to make another visit, this time with pancakes in my backpack. I was glad, too, to find a moment for sitting before the taxi came. It was nine o'clock, sunny, cool and breezy and a wood pigeon or two were calling. Checking the time is not normally permitted but I made an exception today: call it an experiment in the management of train fever. A moorhen called from the dyke.

A large dragonfly flew by, silhouetted, no colours visible, so I thought it more polite not to speculate on species. It was huge: getting towards the top end of possibilities for a modern insect. Emperor? I admired its jerky, erratic course.

A meadow brown flew by, showing off its own version of erratic flight. It comes down to a proper understanding of randomness. You find the same thing in top tennis players: they have the ability to vary the serve – wide, down the T, into the body – in a manner that is genuinely random and therefore extremely hard to predict. Lesser players fall into a rhythm that can be read.

Here were two insects that understood the true meaning of randomness: and it has enabled them to survive across millions of years. I could hear swallows, couldn't see them. Lord, is that the time? I must fly.

> The shuffle option on the original iPod served up songs on a genuinely random basis – and people hated it. You often got clusters of songs together by the same artist. What people actually wanted was something that *felt* random.

Friday 6th

I was back from London, my father frail but cheerful. He had enjoyed the pancakes, washed down with Sauvignon blanc; we had had good conversations, which I had washed down with Famous Grouse. I had also had lunch with Chris, companion on many African adventures, founder and chief exec of the travel company Wildlife Worldwide: perhaps the worst possible business to be running in the time of Covid. He was hanging on, just, on a mixture of bleak practicality and tough-minded optimism. He estimated the chances of our making this year's Sacred Combe Safari at 1 per cent. Even that was something to luxuriate in.

Now as I sat many thousands of miles from the Luangwa Valley, the sky had a bit of everything: high white cloud, low dark cloud, a patch of blue. A kestrel fizzed over, getting the most from a strong southerly wind: if this was a young bird it was doing damn well. The ash tree was making ominous croaks and creaks and I could hardly see the lone alder, the crown just visible: had we ever had such thick vegetation on the marsh?

A kestrel, perhaps the same one, flew fast over my head; a few moments later another kestrel rose from the middle of the marsh. How grand it must be for a young falcon to ride the big wind and do it well, perhaps for the first time: crown prince of the gusting world: air apparent.

The sky was blackening. A dragonfly narrowly missed my head: was that extremely good flying or extremely bad?

> My father always said that young kestrels in action reminded him of the opening of *West Side Story*: as if they were singing, 'When you're a kes you're a kes.'

Sunday 8th

It was raining with some confidence, and the buzzard's calls seemed hungrier than ever. It was gone four and a red admiral flew past, making the best of things in an altogether admirable way. On the Flood the greylag geese were involved in noisy discussions, mostly based on the issue of who defers to whom.

Five mallards flew over: no disputatious troika but a business-like search for food and shelter. Gulls started to arrive in straggles, individuals and loose groups of a hundred or so, coming in from the south and heading for the Flood, where they would spend the night.

Six lapwings flew past: a welcome sight. They had not bred nearby: I would have heard them, for their display calls carry for a serious distance. A wren gave a series of violent alarm calls: what had it seen? This is a good place for stoats. Five greylag geese flew over at 20 feet: close enough to feel the real power of these birds.

The rain was stopping: might as well go in, then, I told myself, only half-joking. But I sat on: and suddenly there was a sound like all the geese in the world. I wheeled around: twenty flying low over the roof of the house, with another thirty close behind, all honking hard. They made a loop over the rookery and came back towards the Flood. Early days of winter, at least for them: things will settle down as they get used to each other again.

> A wren's alarm call is an impossibly loud *tick*; you'd think it must have come from a much bigger bird. It warns other birds of danger: what's more, it tells the dangerous creature – the stoat – that it's been spotted and hunting by stealth is no longer an option.

Monday 9th

I stepped out into the rain to the cooing of wood pigeons and the hunger-calls of buzzards: the days had acquired a certain pattern. A brief shout from a heron.

Comings and goings: a young magpie, a flight of eight mallards, fifty-odd distant crows: a feeling of the world going on about its business. A black-headed gull along the river resisted all my attempts to turn it into a common tern.

The rain was easing: and then, as if to make up for the lack of terns, a flock of twenty lapwings flew over: autumnal birds on the move. Lapwings are in sharp decline: I felt joy and sadness at the same time as they passed. But that's true of many things in the wild world: like having a head for heights, you need a head for sadness if you're to cope with the joys of wild things.

Across the river there was an oozing flock of starlings: 120 of them making a sort of murmurette. The sun came out, a little owl called as if in reproach and then all at once the marsh was full of small whites. A red admiral basked close enough to touch. It may not be the stag and the bear, like Sir Purun Dass, but I was making progress.

> Herons give a loud, monosyllabic call: as a result they are traditionally known in Suffolk by the onomatopoeic name of frank; in the Fens, frank hanser or diddleton frank. Disturb one and it will fly off – and give its alarm call about twenty seconds later: fraaank!

Tuesday 10th

I sat down at half-past five after a nice paddle on the river and a solid session of cookery, for we had guests that evening. It was sunny and breezy, wood pigeons were calling and there was a small cloud of gnats above my head, each male within the group constantly ascending and descending.

I was going to have to move my chair. The wildflower areas of the garden needed their annual mow. It seemed to me important that I put it back in exactly – but *exactly* the same place. A clump of fluffy seeds floated past: the wildflower areas had certainly excelled in the thistle department.

A dragonfly was flying by, eluding both predators and my attempts at ID. This was a territorial male on patrol, approaching me in a rapid zigzag and then performing that dragonfly 180 so fast you'd swear he had a head at both ends.

Just below the lowest branches of the ash I picked out a chunky hoverfly: easy to focus on that one, since he was holding dead still in a perfect hover. This is another form of territorial behaviour or personal advertisement: but as soon as I got him pin-sharp he was gone.

You can only hold still like this if you are a good flier, but that makes you vulnerable, so you also need to be able to accelerate away in an unpredictable direction at the first hint of danger. In this combination of movement and stillness the insect is making a statement: this is a damn good hoverfly and the one you want to choose if you want to make the best eggs.

> The dance of the gnats was also about sex. The territory each individual male holds is notional, useless for practical purposes like feeding and egg-laying, but when a female visits it shows which insects stands out from the crowd.

Wednesday 11th

The rock-steady moan of motor was unceasing. I stepped out into the din because there was no point in waiting for a quiet patch: this was harvest; they won't stop the combine harvester until it's all in and that will take them deep into the night. Summer ends now: first words of Gerard Manley Hopkins' ecstatic poem 'Hurrahing in Harvest'.

I looked up to see if the sky was demonstrating the lovely behaviour of silk-sack clouds. It was, of course, and the garden was almost overwhelmed with butterflies. I bet Hopkins wrote poems about butterflies, but if he did, they haven't survived. Not for the first time I cursed the interfering fellow priest who burned the papers Hopkins left around after his death. A world-beater of a butterfly poem went up that damned chimney.

I found the four sockets for my chair-legs without difficulty, replaced the chair with precision and sat, the smell of cut vegetation filling my nostrils. A buzzard made a hunger-call glissando, reminding me of a grey hornbill, whose song I have a 1 per cent chance of hearing this year.

A blackbird gave a loud alarm call at some unseen enemy – the first time I have heard a blackbird's voice for a few weeks. To my left I could hear the flocking calls of long-tailed tits, a winter sound if ever there was one.

The harvest rumbled on. Could anyone experience an ecstasy of love for the natural world after witnessing modern farming in action? The Hopkins poem was written in 1877. No doubt to a farmer a good combine is barbarous in beauty.

> Modern intensive farming is responsible for many drastic declines in nature. Don't blame the farmers: our neighbours are fine people. It's national policy that's at fault. Organizations like RSPB and your county wildlife trust campaign for wiser allocation of resources.

Thursday 12th

I was off to London again to visit my father, but as usual I found a moment before the taxi to sit for a while. He had been in lowish spirits, unusual for him, when we talked the previous evening, though rallying as conversation continued. I was a bit worried: it was good to sit still for a moment on a day that was cool, cloudy and breezy. A green woodpecker called as I took my place. A buzzard hunger-called – and then paused. This bird was, with great reluctance, beginning to come to terms with the idea of independence: life without parents.

A great tit churred, a contact call: after a few months of near silence while the breeding season was at its height, with a territory established and many chicks to feed, they were making themselves heard again. A green woodpecker called from the middle of the marsh: there are open dry areas where a bird can forage for ants.

I felt rather exposed in my place, with the vegetation so efficiently strimmed down. A stock dove gave a good, sturdy series of calls, just to show that there is more to life than wood pigeons.

I could hear the geese on the Flood: like a wild goose myself I would have to leave this place again, but I would be back the following day. That was a good thought to hold onto as I stood up and walked in the general direction of luggage and taxi.

> As autumn marches on, great tits (and other members of the family) join together in flocks and keep in touch with each other by calling. Great tits make many different calls: the most characteristic is usually described as a *chur*.

Friday 20th

Covid had kept me at home for nearly two years: now for eight days it kept me away from home, the longest I had been absent since my last trip to Zambia twenty-three months back. I was away because my father got Covid and went into hospital.

Cindy had joined me in London for several days as things got bad. She and I set off back to Norfolk in mid-afternoon and I took my seat at seven, both relieved and troubled to be back. My father's condition was, we were told, stable. We weren't, of course, allowed to visit. They would tell us if the situation changed. There was nothing I could do but phone for updates and look out at this prospect, the same as ever and mysteriously different ... when we try to come to terms with complex things we instinctively seek paradox, finding that the truth never lies in the middle but at both ends simultaneously.

It was clear and cool and still, and the wood pigeons were calling. Four ducks completed a complicated manoeuvre over the river. The bright flowers on the great willowherb shone out against the lush green on the far bank of the dyke; from its fastness a moorhen called.

Coming back after a week was a bit like a stop-frame animation sequence of an opening flower or a shifting cloudscape: a sense of change slightly divorced from reality. A buzzard hunger-called just twice; a chaffinch gave one of those *sweet* – onomatopoeia rather than description – contact calls.

As I sat there the marsh was slowly and subtly allowing me back in. A kestrel flew along the boundary dyke; the pallor of the blue sky was offset by complex purples that tinged the few clouds; a greenfinch went through a few sweet phrases – description not onomatopoeia.

Sitting still made me aware of the troubled nature of things: there were no urgent jobs to obscure the hard truths.

I sat there thinking of my father's isolation, of his bewilderment on the few occasions I had managed to reach him on the phone: 'We *can't* come and visit. We're not allowed. You're on a Covid ward.' 'Yes, I know – when are you coming to visit?'

I felt sadder than I had been in London, much sadder, but also a little easier in my mind. I made no attempt whatsoever to unravel this paradox. I stood and turned towards the house: a hare regarded me gravely for a while and then loped away without panic.

> There has been a good deal of research on the value of nature as therapy in all kinds of trying circumstances. I should perhaps point out here for future researchers that nature made me a good deal sadder on this occasion – and I was very glad that it did.

Saturday 21st

It was early afternoon. I had done nothing all day and felt exhausted. It was cloudy, cool and still as I took my seat. I felt overwhelmed and ready for bed, or at least for my hammock, a birthday present that was now strung from the ash.

A sky-blackening great black-backed gull flew over: they are colossal things with a five-foot wingspan and, when close, quite unmistakable. A buzzard flew easily over the distant oaks; there was a hint of goldfinch call and a clear great tit. It was reassuring how little had changed during my absence, but then it wasn't the season for abrupt changes and new arrivals.

A tight group of five ducks flew over the river, teal I guessed – tiny ducks with whirring wings. Another bird was following the river in the opposite direction: wood pigeon, I thought, lifting the bins lazily – and of course it was a hobby, purposeful flight on needle-pointed wings. I appreciated the gift of a small marvel and smiled a sort of general thanks to the marsh and the forces of nature while two wood pigeons started calling, one in each ear.

A twittering overhead: swallows, but where? Over they came, eight of them heading east: not hawking seriously or flying anywhere with purpose: a schooling outing, learning the nature of flight. Soon they would be flying to Africa: take me with you! But no. I must stay here. Even if I was allowed to go, going is impossible. I have things to do. I went across to my hammock and took a swinging doze.

> 🍃 I was tending to make these calls to the hospital outside. For reasons I had no wish to analyse, it made these grim things more easy to bear.

Sunday 22nd

It was cool, a gusty breeze and occasional bursts of sun: a buzzard hunger-called and a deer was baying as if he was the fiercest thing in the world. Then 300 starlings, conservative estimate, rose above the horizon, made a classic amoeba shape and dropped back down again: in sight for just five seconds.

The young buzzard flew over the marsh, still calling, looking as if he needed stabilizers. But you're almost on your own now, so it's sink or fly. Along the river I picked out the pale head of a female harrier; behind her the restless honking of the restless geese.

Another buzzard, an adult, flew across the marsh, wings flat as a table-top, no hint of a dihedral, and settled in the lone alder. No harrier came to harry it: perhaps the season for such aggression was over. Must be a nice change. All at once forty ducks leapt up from the Flood: as I was wondering why, a male harrier – a perfect male harrier – flew by. I felt for an instant that I had the meaning of all things before me, only just out of reach.

Two butterflies rose in a nuptial flight: against the light they looked simply black: peacocks or red admirals, probably. Another wood pigeon, but I raised the bins anyway. It was another falcon. Talk about being a bad birdwatcher. In a moment of intensely private connectivity I watched the young kestrel powering on towards its destiny.

> It's always worth having a proper look. It might be something thrilling of course, and it might be a nice view of something ordinary. When you're trying to understand time and place and nature, both experiences are good. It's also pretty good when you're trying to come to terms with terrible things.

Monday 23rd

First, get on with the job of worrying. Then seek scanty information from the overworked hospital staff. After that, pass on the news to my sisters. I had that clear in my mind so, just to show how well I was coping, I picked a fight with my computer, which kept defaulting to United States English. After a morning of rage, punctuated by color and maneuver and the distance in meters, I fled to my seat to let the breezes mend at least some of the bruises I had administered to my spirit.

There were big round bales on the most distant field, the harvest done, the plough to follow soon enough. A sound of splashing and honking wafted over from the Flood and, perhaps related, peevish hooting from a boat on the river.

Even the wood pigeons were quiet. There was just the occasional call from the young buzzards. I wondered where the cuckoos were now, and whether they'd be back next season. Who teaches a baby cuckoo to migrate? Certainly not its step-parents.

A disturbance to the vision of my left eye: I made an adjustment and saw a spider hanging from the peak of my cap: not my favourite taxon, but this one was tiny and easily manageable. It was a balloonist: spiders can travel miles on lengths of silken thread, making a landfall or capfall wherever chance takes them. I unattached the silk – the cap bore the legend Remote Africa Safaris – and set the spider gently earthwards to seek its fortune.

An adult buzzard called: I found it, and then a second bird flying beneath. The two climbed together in circles and then one attacked the other with outstretched talons. But there was no aggressive response from the second bird; perhaps there was some kind of flying lesson going on.

I wouldn't claim a conscious intention to impart knowledge, though I wouldn't rule it out either. And perhaps there was a conscious intention to learn. How can I become a writer? A question I've often been asked. Read, I always say.

Read, imitate, respond to what's around you and eventually you will work out how to do it for yourself and in your own way. The two birds rose together, one I think imitating the other, riding a slim thermal; behind them I found two more birds, both impossibly high and both hobbies.

> 🍃 Grove, the local cuckoo satellite-tagged by the BTO, had fallen silent, they reported. He had perhaps been predated, or his tag might have failed. He at least survives as a minute chunk of data in the research into the decline of cuckoos.

Tuesday 24th

I thought I was seeking quiet. Certainly I found it: it was as if the countryside has fallen asleep: no movement but the wind. It was intermittently sunny, the sky studded with bulbous cumulus clouds. A young harrier moved briefly along the river: above me a great black-backed gull.

And I realized that I didn't want quiet at all. I wanted distraction, sensation, huge exhilarating sights I could boast about for years. A buzzard crossed from the heronry and vanished into the lone willow, apparently gobbled up like the hobbits when they met Old Man Willow. I could hear a goldfinch.

That's when I broke the iron rule of sitting and reached for my phone. I had been trying the ward all day and failed to get through. This time I was lucky: a doctor gave me a long, lucid update. Nothing to make me get out of my seat and cheer, but it was clear things could be a great deal worse. The ward was run by good people doing their best in appallingly difficult circumstances: here was someone able to give me a little time and thought, and I was very grateful. So I got up to pass the news around the rest of the family.

I went back again later and sat, but it wasn't easy. I needed to be busy, not still. I wanted to be hectic, not calm. I sat restlessly as a restless mob of birds rose from the Flood: gulls and ducks but among them a few lapwings. It's always good to see lapwings.

> I remember seeing huge flocks of lapwings as a boy and taking them utterly for granted: just peewits. They have made the dreaded transition from ambient birds to special birds.

Wednesday 25th

The 1 per cent chance had gone: I wasn't going to Africa, Chris told me. So it goes. Not that I'd have gone. Something that had mattered so much was now a footnote. It was good to be sitting here, on a cloudy day with a fair bit of breeze: hunger-calls of buzzard, hooting from a moorhen, quacking from the Flood.

A buzzard crossed, heading away from me, silhouetted, giving me the chance to admire the considerable surface area of the wings: nothing dainty about a buzzard. It was perhaps a young one gaining mastery of the place beneath and the air all around. There was an outbreak of honking and wing-flapping from the geese on the Flood: things seemed no calmer out there. As the sound died away I seemed to be sinking deeper into my chair, the chair sinking deeper into the earth. Today the stillness and quiet were good.

In this moment of rare comfort, a bird, surely a pigeon, crossed rapidly and in my stillness I was half-inclined to let it go. But I raised my glasses conscientiously and found a hobby. Then at once the sky became a traffic jam: 300 gulls flew over my head, bearing left for the Flood, taking five minutes to do so. Basho felt the autumn wind in the taste of a bitter radish: I could feel it in the flight of these flocking gulls.

Basho, the great eighteenth-century haiku master, wrote:

In the bitter radish that
bites into me, I feel the
autumn wind

> Haiku will help anyone who wants to 'get' nature. Best introduction: *The Classic Tradition of Haiku*, ed. Faubion Bowers.

Thursday 26th

It was another bitter-radish day: cold and cloudy with sudden hold-your-hat gusts. A buzzard, probably a young one, crossed the marsh, side-slipping with the wind: I felt I should award marks, like a judge at skating or gymnastics. Not bad, but not a perfect ten.

More sky. I wanted more sky. Perhaps I would lop down some of the blackthorn that was hiding so much sky. A long-tailed tit called, just once. Once again I seemed to be sinking into the earth: was this illusion? I checked: the chair-legs were now in sockets two inches deep. Would it be acceptable to raise it up?

Two buzzards appeared from behind the heronry, one with a distinct wobble. Was that a failure to read the wind and pre-empt the gust? Or a failure of technique as the gust hit? The wind is not to be mastered, as any yachtie will tell you: all you can do is work out how best to cooperate with it.

Today all the passing falcons were pigeons. What shall I do when my year is complete? I felt a bitter-radish chill at this thought: both at the passing of time and at the sadness that comes with the completion of any significant project. But I will sit on, will I not? On a day full of questions, here was some kind of answer.

> This was not, as I'm sure, dear reader, you will understand, a time when you make a point of counting your blessings. All the same I was very much aware of them. They included easy access to a wonderful bit of nature – and even more than that, the *habit* of nature. If you, dear reader, ever find buzzards on a day of sadness, this book will have been worthwhile.

Friday 27th

It was warmer, stiller, sunnier. What's more there was better news from the hospital. It was remarkable how my anxieties expressed themselves as physical aches and pains. I sat alone feeling a powerful kinship with everyone else who has had similar troubles in these troubled times.

Geese and ducks called from the Flood. I admired the great willowherb and allowed myself a small moment of self-congratulation: at least I could now tell it from the equally gorgeous but alarmingly invasive Himalayan balsam. Conservation volunteers on the Broads make up weekend parties to go balsam-bashing. Balsam is like Nicole Kidman in *Paddington*, I thought: beautiful and dangerous.

A red admiral perched head-down on a nettle stem and I focused on it. Did you know that red admirals have a red band on the trailing edge of their hindwings? And that it's beaded with black? Once again a revelation of the marvellous ordinary: Basho would have caught the moment in a perfect haiku.

A buzzard got up; a great tit was calling in the ash ... and then a flash of shimmering green and a damselfly passed six feet in front of me to perch on the nettle the red admiral had just vacated: a slim green needle bearing four wings, all folded slightly back. As I stood I heard a little owl.

> On returning to my hut I opened *Britain's Dragonflies* at random: the page before me showed an emerald damselfly; the text told me they were on the wing in late summer. Life should be like that.

Sunday 29th

The news from the hospital was cheering: he was not only a little better, he would soon be moving to a non-Covid ward that allowed visitors. I sat out in wary good humour, on a cool windy day with high cloud and occasional moments of sun. I looked with pleasure towards the heronry: I had given myself an extra ten degrees of view by lopping the blackthorn.

I had intended to be entirely passive in this venture, taking only the adventures that Aslan sent me. But the end of summer brings a mood of restlessness and taste for new projects: perhaps that's why the academic year begins in September.

There was much quacking and honking from the Flood: a dragonfly came with a flash of blue and yellow, possible Southern hawker. The wind kept swirling: earlier, as I had paddled along the river, I had faced a headwind in both directions, or so it seemed. Still full of reforming zeal I stood and lifted my chair from its deep sockets and moved it a few inches. I sat again and felt almost dizzy with the added height, the wider view –

– and all at once the sky was full of birds of prey, four buzzards and two harriers, the buzzards calling as all six birds rose in a spiral together. Two buzzards flashed talons at each other, in play rather than mortal aggression; below them the harriers climbed in a neater, more economical style, though the buzzards climbed higher and faster.

If there was enough life down there on the ground to feed these six sky-blackening marvels then the landscape before me was doing all right. For the first time in weeks, I wanted to cheer.

> 🍃 Ecologists talk about carrying capacity: the numbers that an ecosystem can support: plants to feed vegetarians, small creatures to feed large creatures: and if the top predators are surviving, things could be a good deal worse.

Monday 30th

My father would enjoy sitting here, I thought, if he could be beamed up, if he could be wrapped up, if he could manage the sitting-up bit. There had been more good news, more talk of the possibility of visits. The geese were working up a crescendo on the Flood. It was still cool and overcast, and as a fierce gust hit I watched a dragonfly duck into shelter. The air was full of the calls of what James Joyce called peasants and phartridges.

Two swans flew along the river: the first I had seen – and heard – in flight for many weeks. When food is plentiful and there are cygnets to be raised, swans don't need to fly much; it's not as if they have many enemies to flee. Now there were swans on the wing again. You always notice firsts, never lasts. A young buzzard called, perhaps just out of habit, and fanned its tail so wide it almost met the trailing edges of its wings, so that the birds seemed to be one huge wing.

I found some movement very high, focused and found half a dozen house martins. They were moving north in a leisurely way. So not thinking about Africa just yet. All the same, they were perhaps the last I would see this year.

> My father loved walking more than sitting, always with a collie – Floss, Lettie, Bess – at his heel. But sitting is one of the pleasures of any decent walk: you find a nice place and sit (waterproof trousers or supermarket plastic bag for wet days). It feels like a well-earned reward and that makes the landscape before you and those who live in it especially pleasing.

Tuesday 31st

It was gone five and my sit felt like a well-earned reward after a paddle on the river. I was cheered – cautiously – by still-okay news from the hospital. The air was full of the yelps and honks of geese on the Flood: the quacking of female mallards sounded like complaints about noisy neighbours.

A young buzzard called; a few jackdaws were chattering at each other and a moorhen called sharply from the dyke. And then came the swallows. I first picked them out quite distant, visible only as specks that swooped in the manner of swallows, and then as birds that clearly possessed the tails of swallow (not martins): and above them, almost impossibly small, a harrier.

As I watched I found more and more swallows, a good fifty of them, swerving and looping. They looked quite tireless, as if no distance could ever be a problem to them, still less gravity. The harrier was still there above them, turning and turning.

And then ten geese flew over at 20 feet, occasionally

calling to each other. They made a quick economical circuit before heading to the Flood.

> 🍃 A swallow's outer tail feathers – the longest – make constant tiny adjustments as the bird manoeuvres: engineering that humans can't recreate. It's worth noting that the word aviation comes from Latin and means, roughly, the stuff that birds do.

SEPTEMBER

Wednesday 1st

I couldn't get through to the ward: I would try again later. I sat there remembering the almost euphoric days of early Covid eighteen months back, a nation united in the resolve that we *would* get through this thing. Accordingly I sent a round of silent applause to the NHS, especially the people on my father's ward, especially the nurse with the beautiful Jamaican voice who called me darlin' and told me how great my father was. It was cloudy but bright, and I was surprised by half a dozen – no, more – ten lapwings heading west.

A long-tailed tit was calling animatedly from the willows; the honking from the Flood had settled to a mellow murmur. A shotgun boomed in the distance: it was September and the time of gunfire had returned. So it goes.

And then the sky was filled with lapwings, a great loose flock that filled the air above the marsh, a good 100 of them: a lapwing century, and, it seemed, the best possible omen. They moved about on their big round wings, so unhandily floppy, pale underneath and dark on top, colouration that stressed the rhythm of their flapping and made their passage seem like a ballet of the airways: gallant aviators sent here to bring comfort and joy. The flock split: most of them went east and a dozen or so went in the opposite direction. But after a while, this minority party changed its mind and came back in search of the rest.

A little later they told me my father was on the move as well: migrating to a different hospital and a non-Covid ward where visitors were welcome. London tomorrow, then.

> Lapwings give great flying displays in autumn and winter when they gather in flocks. You can find the best spots to see them by checking the websites of the RSPB and your local county wildlife trust. You can also find them at the London Wetland Centre.

Thursday 2nd

Have you ever travelled in a balloon? If so you will have experienced – in between fierce blasts on the burner that keeps you aloft – an acute and deafening stillness. You are moving at the same pace as the wind, so there is no air moving past your ears: as a result every sound is so clear it seems magnified. That feeling was replicated on this morning of perfect windlessness. It was just gone nine and I had found time for a brief sit before moving towards London.

Sounds came at me from all sides: caw, quack, coo, quirk and honk. Two wood pigeons exploded from the ash in a delayed response to my arrival, though I carried no shotgun. A flock of thirty mallards got up from the Flood and flew along the river before turning back. I could hear a distant train beyond the valley.

A herring gull called and was followed by a lesser-black-backed gull: like a recording specially designed to demonstrate the difference. Townies would be complaining about the din: get back to London for some peace. Well, that was exactly what I was planning.

> The Natural History Museum website has a terrific section for birdsong beginners, giving you the ten commonest to listen to. Search for 'nhm discover birdsong'.

Saturday 4th

Eyes. They reveal the important things, in those we know most intimately and in complete strangers. They also reveal things about our fellow mammals; I know this best from horses, but also from dogs and cats, and on occasions – one or two of them heart-stopping – in the wild.

And my father's eyes were good: lighting up when he saw me, full of life as we talked. He was over Covid: they were working on recuperation and a return home. For this relief much thanks: a line, of course, from the first scene in *Hamlet*, my father's favourite play. So we talked about family and friends and incidents from the past and books and walks and birds we had shared. Then I read him the opening scene from *Hamlet*. The next day we did it all over again: the peregrine at Trewavas, the seal at Stackhouse, the mnemonic for the song of Cetti's warbler ...

The day after that there was a Covid case on the ward and the place was closed to visitors for the next fortnight: more isolation was the last thing he needed. So I was back in Norfolk and so was the breeze, and I feared the fitful sun might get blown away. A great spotted woodpecker called; a crow gave four powerful caws, paused, and then six more. A dragonfly came fizzing down the dyke with the wind behind it, as if in the most terrible hurry to get everything done before autumn.

I looked up and caught the neat silhouette of a soaring sparrowhawk, circling twice, gaining height, sleek perfection in every movement. As I watched I found a buzzard still higher, maybe 400 feet, and a little below it a red kite descending in a swift, shallow glide before turning crosswind, flapping those long, bouncy wings, the forked tail twisted almost painfully into the turn, the sun picking out the red bits and making the bird shine in a way that seemed wholly exotic. The kite and the buzzard passed each other close, each ignoring the other.

The kite climbed again, now very high indeed, with two buzzards just below; one of these assumed the classic W shape and descended in a steep, purposeful dive; the other stayed high, still climbing, tail fanned out hugely. In the far distance I picked out a single swallow.

> Here's another classic difference of design. The kite's long forked tail is good for manoeuvrability, especially manoeuvrability in the glide; the buzzard's broad fan-tail is good for soaring, for gaining height without flapping.

Sunday 5th

There is comfort in routine as well as in nature. Perhaps it's a double comfort when nature is part of your routine: as I stepped out I saw a sunlit bum, neat and furry and clearly belonging to a deer. I held still: after a while the deer moved and turned a small sweet face towards me, large-eared, large-eyed, watching me in wonder. I stayed as I was, not quite invisible under the deer's stare. It was pretty young, its experience of the world measured in months. We held the stand-off for two or three minutes until the deer broke it, but even then without panic: instead a sedate canter into cover.

I took my seat beneath a sky of pure blue, a little real

warmth and a gentle breeze: a day anyone might choose to sit in the garden. Two swans flew along the river, and then another two.

The air above the vegetation was full of insects: busy busy busy: gnats, dragonflies, butterflies, hoverflies: the sun brings energy and the whole marsh was revelling in it. A big dragonfly buzzed me twice, as if for a dare, the second time so close I jerked my head out of the way.

Here was largesse, though only a single swallow was taking advantage of it, heading northeast in a series of easy curves. Beyond the river there was a sudden snowstorm of gulls, a good hundred of them – and just as suddenly they were gone.

> Most insects get their heat from the environment, rather than generating themselves as we mammals do. Warmth allows insects to be more active: so a nice sunny day is best when you're looking for butterflies.

Monday 6th

It was sunny again and there were small whites on the wing all around. A young buzzard called, and then an adult. The return to the routine of telephoning the ward was an unhappy business; but at least I was now sitting out with the job done, the news equivocal.

I could hear the sounds of a creature feeding in the dyke and told myself not to disturb it: just listen to the splattering and the pabbling. I guessed a filter-feeding mallard. A small fly collided with my face: there were a lot of insects on the wing. The long-tailed tits were calling with increasing energy: it's a sound that gives you good cheer on a winter day, and they were already building their busy little flocks. From the dyke there was a sound like a belly rumble, followed by a derisive snort.

Four swallows crossed in loops, heading east and then turning south. A small, drab moth fluttered past me and landed on a grass-stalk: with immense care I dropped to my knees and found a creature of exquisite beauty: sumptuous patterns in subtly differing shades of green. I took a photograph: it turned out to be a green carpet moth. It's easy to miss such beauty. But today I didn't.

> Mallards do a fair bit of filter-feeding, like the blue whale only different. They take water into their beaks and push it out with busy tongues through serrated plates on the inside of their beaks; the dykes are rich with aptly named duckweed.

Thursday 9th

Three days had passed. I sat in the unexpected sun, warmth soaking into my bones, butterflies all around. It was good to be sitting still. It was good to have a still mind, noting, rather absently, a call from a green woodpecker. I felt an eerie gratitude for everything that lay before me: grass and nettles at my feet, the ash above, the reeds and sallows of the marsh, the flying insects that were like stars at dusk: the longer you look the more you see.

My father had died three days ago. He'd got over Covid but had nothing left to deal with what followed. I had gone to London, got there too late, not that it would have mattered to him, for he was heavily drugged at the last. My sisters and I gathered together and gave each other such comfort as we could. Then we got to work on the mundane practicalities of death.

Now I was back in Norfolk, taking comfort and meaning from the same unchanging ever-changing prospect before me. My father was not a contemplative man: not a great sitter, until increasing age gave him no other option. He loved nature and appreciated it all his life, mostly on long dog walks with binoculars round his neck. I thought of our week-long walks along the Cornish and Suffolk coasts, sharing avocets, choughs, marsh harriers and once a 12-foot basking shark.

He introduced me to birds, but I paid him back in later life when I taught him how to get in deeper. I had shown him many birds and bird-places and birdsongs. Nature had been a great shared thing in our lives. The thought pleased me as I sat on. The sudden chatter of a group of magpies broke the silence: as noisy as a Barnes family gathering.

This was a butterfly day: small whites and red admirals: the warmth was an opportunity that must be seized if more butterflies are to be made. The Greek word for butterfly is *psyche*, which also means soul. In a good sadness I sat on, surrounded by butterflies.

> I hadn't really planned it that way, but my father turned out to be a major character in my book *How to Be a Bad Birdwatcher*. As a result I dedicated it to him: 'The first bad birdwatcher I ever met. He taught me all he knew.'

Friday 10th

It was warm, overcast and breezy, a few gulls over the river, a few corvids beyond. The news had got out and there were obituaries in the papers. My father had worked for BBC Children's programmes, producing *Blue Peter* and then revolutionizing the entire concept of television for children with his invention, among many other programmes, of *Swap Shop*, *Grange Hill*, *Record Breakers* and *Newsround*; for the last he was given a Bafta. There were also many cheering messages of condolence (my father always called them 'condoms'). The family's mutual support system was holding up well. A dragonfly scurried along the dyke in front of me.

Then a gorgeous female harrier made a low run over the marsh, large, elegant, pale head clear as if picked out by a spotlight. I had introduced marsh harriers to my father, explaining their double bounce-back from extinction, showing him the dihedral. It was a payback: when I was ten he had introduced me to buzzards, then something of a rarity, sighting one over the Fal estuary. We had shared many wild times and wild sights.

I saw a distant eagle in my peripheral vision and turned my head sharply: it turned out to be a very small hoverfly a few yards away. The illusion of distance is the subject of

Edgar Allan Poe's story *The Sphinx*, in which the narrator perceives a terrifying monster on a distant hillside, later revealed as an insect very close: the ultimate expression of the Dougal Principle.

The sun came out and called forth a little snowstorm of small whites over the marsh. A jay called like a suddenly furious cat, reminding me of my father's occasional volcanic intemperance.

A small dragonfly perched briefly on the peak of my cap, made a circuit of my still self and then alighted contentedly on my sleeve to soak up the sun's warmth: as the Buddhist said to the hot-dog seller, make me one with everything. It had a brown body a couple of inches long, and a small dark spot near the tip of each of the four wings: A female common darter, I later learned. A gust of wind, and a little bunch of leaves fell from the ash, yellowed and done.

> I have shown wildlife to many people besides my father, in books, in Africa, in many other places, and it's always gratifying when people tell me how wonderful I am. But alas, it's not me that's wonderful: it's what I'm pointing at.

Sunday 12th

A savage thunderstorm had overwhelmed us on Friday evening: now it was sunny and cool, as if the storm had never happened. There was more pabbling in the dyke: earlier research had revealed these pabblers as two swans, like the mallards filter-feeders, but on a more massive scale. I could just make out their white forms through the stems of nettles. These were beginning to die back, yellowing, and even some blackening leaves. A wren ticked sharply; one of the swans made castanets of its beak.

I sat back. Different tones of buzzing filled the air, which was rich with insects. I knew why a butterfly is called a butterfly, but why is a dragonfly a dragonfly? I made a note to look it up. Overhead I picked out a small wader moving at top speed, at a confident guess common sandpiper.

Once you start looking at the sky you get sucked in, or drawn up, rather: there was a crow, and higher still, much higher, there was another bird with fanned tail and spread wings. It was then that I did something absurd: remembering again my father's Fal buzzard, I saw this buzzard high over the marsh as – oh, I don't know, spelling it out rather takes away the moment and the meaning, but the buzzard was, for a moment, my father's ascending soul or something.

From the dyke a swan made a sort of bassoon noise and then flapped his wings a couple of times.

> Butterflies are named for the gorgeous brimstone, which is bright yellow, the colour of butter. Francis Bacon used the term dragonfly in 1626, the first recorded instance, and probably borrowed it from a folk-name. He might have chosen instead devil's riding horse or devil's darning needle: sinister associations for an insect that does no harm. This from Peter Marren's excellent *Bugs Britannica*.

Monday 13th

It was sunny but cool, a sharp wind from the northeast: not a day for visions of ascending souls. The young moorhens were making squeaky calls from the dyke; a dozen yards further on the swans were still in residence. Small whites fluttered over the marsh.

Another wood pigeon flew over. After a pause, waiting for something more thrilling, I made a note about it. At the end of play the day's most successful England cricketer gives a brief press conference, but on days when no one excelled, the then head coach Duncan Fletcher would turn up instead. In the press box a thoroughly rotten performance was always a Duncan Day.

Well, this was a Pigeon Day. A crow called; there was a quick volley from the Bauhaus. A flock of twenty finches flew across, bouncing in the air. Then a female harrier crossed the marsh and my spirits lifted a little and carried on rising, very slightly, in an elegant dihedral. A dozen swallows came hurrying past, feeding as they went, heading a little south of east. I sat on; the swallows looped back, finding food above the marsh, no hurry to leave just yet.

> Swallows don't need to fatten up before migration: their food is in the air, so they can feed as they go. They travel by daylight for that reason, covering about 200 miles day, resting up at night in large flocks at traditional stopover points.

Tuesday 14th

It was my mother's habit to say, as she sipped the day's first whisky at precisely seven o'clock: 'I can feel it doing me good.' As I took my seat I felt much the same thing, though a later whisky would no doubt continue the good work that nature began. The rain only added to the goodness.

The moment after supper was a daily sadness, for that was the time I always called my father, every night until he went into hospital with Covid. Often the conversation would go on for half an hour, recalling lost times and the many and curious sayings of my mother. It was good to think of both my parents as the rain fell.

I hadn't sat in proper rain for some time, and it seemed that I had missed it. I was there in full waterproofs and brimmed hat, picking out a cormorant flying along the river as the rain fell, steady rather than fierce, rattling the leaves above me.

I had my bins at the ready in my lap: a hoverfly, perhaps catching sight of his own reflection in the lens, came in close to investigate, confident his Batesian mimicry would keep him safe. Eight lapwings flew east in a ragged bunch: ragged formation and ragged wings.

A devil's riding horse cantered around the space above my chair: more than any other flying creatures, dragonflies do corners, forever making sharp 90-degree turns. Do they see their territories as squares? Or cubes? Or is it a strategy for dodging predators?

A handful of swallows flew over, feeding and heading southeast. A young moorhen called, a sound like striking a balloon a glancing blow with a wet finger. And then thirty lapwings flew over in a most disorganized flock, apparently in the middle of a furious debate about where they were going. In the end most of them went northeast, while a straggling half-dozen decided their destiny lay to the west. The swallows looped back, curving, slowing, accelerating.

> Black and yellow are warning colours in many different species that really do sting: bumblebees, honeybees and wasps all flash the same signal. The tendency of dangerous creatures to look the same is called Müllerian mimicry: if an enemy is warned off at sight, there's no need to waste energy (and sometime life itself) in a sting.

Wednesday 15th

I took my seat rather wearily: in recent times the only time I hadn't felt like sleep was in bed. There was a spread of patchy cumulus clouds before me, breeze and even a little sun: I wanted to pull the clouds over my head like a duvet and sleep for the next decade. I could hear cheery quacking from the Flood.

A dragonfly flew directly away from me, not in 90-degree corners but in shallow zigzags, say 135 degrees; that must be about predator avoidance. A robin started to sing from the ash, relaxing into a strong, sweet melody. Two conjoined dragonflies passed along the dyke: a pre-mating ritual.

A couple of lapwings headed southeast; four swans flew along the river followed by four more. It was like the game of Happy Families: have you got Mr Cob the mute swan? Not at home: have you got Miss Pen the cygnet? The trailing quartet looked slate-grey in this odd light until they turned left over land and their wings lit up white as angels. The sound of their sixteen wings filled the valley.

I heard a musical whistle from a dog-walker and then remembered there aren't any dog-walkers round here. I heard the sound again: a curlew, the first of the autumn, coming down from the hills to spend their winter with us. I found the bird, flying over at 50 feet, moving at some pace, calling repeatedly – and then I realized it was not alone. It was catching up with the rest of them: a flock of twenty-three already beginning to drop down behind the heronry. Once again I savoured the consolation of ornithology.

> You will often see two dragonflies making an eight-winged tandem of themselves. This is not mating; it's what comes before: the male grasps the female's head with a structure at the base of his tail called claspers, and they fly as one. After that they mate in a sort of soixante-neuf position. The act takes a few seconds in some species; in others a few hours.

Thursday 16th

I had had enough of the day by noon. Three hours later I was sitting out without any great hopes, either of majestic wildlife or comfort and consolation. It was warm, with patchy sun. Way over on the far side of the river was a gathering of gulls; the land had been newly tilled, and the birds had been feeding; now they were resting up.

It was a good day for insects. How can I make that sound attractive, I wonder? There were plenty of wasps and mosquitoes about, few people's favourite living things, many other species as well – and though this is good for a million reasons, it's hard to express that without inciting revulsion. When there are no more insects to feel revulsion for, we'll all be lost. A robin, an insect-feeding bird, sang cautiously; further off a heron called frankly.

The gulls all took to the air, 200 of them; as one they dipped, rose again and then most of them dropped back down. Those that did so were black-headed gulls; a dozen lesser black-backed gulls preferred to stay in the air.

I caught a glimpse of a female harrier, a bird that was both rare and common, exotic and homely, ordinary and marvellous. It would surely be a better day tomorrow. The robin struck up again: proper song this time: a bird that was planning to live through the winter and make more robins when the weather got warm again.

> Flocks of birds, shoals of fish, all apparently controlled by a single mind ... it is one of the beautiful illusions of the wild world, and it comes about because the response-time in small animals – the distance an impulse has to travel from brain to muscle, from intention to action – is rapid beyond easy human perception. What is sequential looks like unison, and that is the joy of such things as starling murmurations.

Friday 17th

Two young harriers were there to greet me as I sat down, promising a golden future for harrierkind. It was a warm day, the sun in and out of cloud, a light breeze. Long-tailed tits called from the willows to my left.

The wind dropped and for a few minutes the warm sun on my shoulders was like a comforting arm. The robin sang from the ash above me: the time for tuning up was long past, this was proper song, song with which the bird would guard a feeding territory throughout the winter and, if all went well, survive to make more robins in the distant spring. From the dyke an unseen swan beat its wings three times in a sort of heraldic yawn.

And then without any drama at all the sky was full of lapwings, forty of them, flying about as if they were common, as if half a century of trouble had never happened. They flew around for some minutes in their unsettled way; eventually most of them headed off east while four of them went south: flock solidarity is not strong among lapwings. Above them but paying them no mind, a buzzard. As I watched I found a harrier a little lower: significant dots, given shape and meaning by the magic of familiarity. And decent optics.

A better day.

> You see best what you know best. That's the way the brain works. It follows that the more familiar you are with wild things, the more clearly and the more often you will see them. Every time you look and listen is an investment, one that means you will see more and see better next time.

Sunday 19th

It was the last few hours of a good weekend. My sisters had both come to visit: the idea was to discuss the funeral and related stuff, and while this was all essential, other things mattered more. We had hired a boat and chugged around the Broads for an afternoon: the watery landscape, the big sky, the on-board treats and the togetherness doing us all kinds of good. After they had gone I sat out at half-past three, the cloud high and the weather warm on this, the last weekend of summer.

More or less at once a small group of swallows flew in and started hunting in curves, as swallows do. They were moving south in principle, but in no great hurry: they had come across a source of abundant food above the marsh and it would be a shame to pass it by. I thought they had gone after ten minutes, but back they came, zipping over at around 20 feet, too busy and spread out for accurate counting, but I guessed around fifty.

They flew as individuals within a group, not as cogs in a machine, not like starlings in a murmuration, each one constantly changing pace and direction without reference to any other, slowing, moving into a glide, then turning and accelerating with a sudden whir that made their wings invisible. Their energetic aerial life needs a lot of fuel, and they were filling up: jink, slow, flat-out charge, then apply the airbrakes and savour the next flying canape.

Summer was almost over, the swallows almost gone. And, I thought with sudden alarm, my year-long journey was almost complete. Well, the years and the earth will keep on going round and, God willing, the swallows will return. And as to future stationary adventures, the matter is in my own hands. Or bum.

> 🍃 Birds not only look different: different species also fly in different ways, ways that suit the niche they occupy. The more you look, the more you are able to tell birds from the patterns they make in flight.

Monday 20th

If you read your old *Blue Peter* annuals again you'll find that each one contains a detective story, one in which the criminal gives himself away by making six very foolish mistakes. Can you spot them? My father and I had worked together on these stories: it was fitting that out here, on this nice sunny day, I made two very foolish mistakes in quick succession.

But first I noticed a fleabane, a very jolly yellow flower, on the edge of the dyke and wondered how long it had been there and how I had missed it. Too busy looking at the sky, I suppose. As I was chiding myself a bird came in sight for about 0.25 seconds. I had an impression of a pointy wingtip and a flash of green, no more –

Bee-eater! I said the word out loud, for I know several African species very well. And before I had time to think my mind was off again: just wait until I tell him on the phone tonight! Two seconds – no, perhaps as much as three – had elapsed since that first flash of green before I was aware that (1) it almost certainly wasn't a bee-eater and (2) tonight my father wouldn't be shouting, 'Bee eater? Bloody hell, boy!'

A pair of bee-eaters had nested that year in Great Yarmouth, of all places. I listened for the distinctive call – and bee-eaters are always calling – *prr-up prr-up*, the nearest field guide tells me. I heard nothing. But I remembered Wordsworth, who was surprised by joy and 'turned to share the transport – Oh! with whom But Thee, long buried in the silent Tomb'.

Above the horizon I could see a flock of birds, oozing and shape-shifting in a pleasingly murmuratious way, but within thirty seconds they were gone, long before I could assign them some fanciful identification.

But here was a fine red kite, catching the sun so perfectly it was almost a scarlet kite – and hang on, was that a bee-eater calling? Or was that wishful hearing? A jay flew over, the light turning it almost orange. I'd better go in before I write it down as King of Saxony bird-of-paradise.

> Bird identification is a skill, not a gift, and like all other skills, it's a combination of natural ability, hard work, practice and the will to improve. A better birder than me would have either confirmed the bee-eater or categorially dismissed it ... well, probably. All birdwatchers are chasing perfection, knowing that perfection is unattainable.

Tuesday 21st

It was another nice day. Carl, my personal rarities committee, had told me that a few bee-eaters had been seen along the east coast of Norfolk, riding the southerly winds, so yesterday's snatched glimpse was – well, let's call it a just-about possible possible. If only I had had it in sight for another quarter-second ...

A fly was buzzing round my hat, but I ignored it – and then wondered why. It must be because flies make a different buzz from bees and wasps. A distant crow was making a dart at a buzzard, but its heart wasn't in the job: the mobbing season was over, the chicks fledged. A collared dove called, mildly unusual for here; they like villages better.

It was a quiet day, then ... and so I found my mind drifting off, and I was composing the eulogy I would deliver at the funeral in a couple of weeks. Well, that wasn't really so inappropriate; I would certainly mention the birding we did together. I followed an altogether appropriate buzzard across the sky. Another, a young one, called: but it was very nearly the sound an adult makes. Moving on, moving on.

> This instinctive awareness of different buzzes reminds me that no one is an absolute beginner when it comes to wildlife. We all have knowledge, often deeply buried, some of it picked up along the way, some of it innate: and it's all ready to help us on when we turn our minds to nature.

Wednesday 22nd

The equinox was to take place at 8.20 that evening: for the next six months the nights would be longer than the days and that's always a solemn thought. Today the air seemed warmer, the marsh more peaceful, all of nature more at ease with itself – though perhaps this was because I had written my eulogy and found it an important step in the processing of grief. I had included jokes, as my father had insisted when planning the event some years back. The jokes were painful to write down, but they seemed to have more meaning than the solemn bits.

Now I was sitting out again as two butterflies, unidentifiable in silhouette, performed a nuptial flight. A jay flew over: jay-time had come around again, the weeks ahead to be devoted to the gathering and caching of acorns. A kite passed by with that nonchalant flap of the wings that kites specialize in. A dragonfly fizzed past me at knee-height, catching the light to perfection so that it was identifiable – even by me – as an emperor, huge and with a big pattern of green and blue.

I was then struck a light blow by a falling frond from the ash, a stalk and three yellow leaves. Despite this obvious bit of dying the air was full of insects and there was life-making going on all around. I had gone out seeking an elegiac mood: but there was far too much life for such self-indulgence. I could also hear Cindy's power tools in action: the reassuring howl of art. Beauty would come from this din.

A quacking filled the air. I wrote two words in my notebook: Distant Ducks: just like that, with initial caps. Then I wondered why ... until I remembered my father walking the cliff-path in Cornwall to meet family members for a pub lunch. He was late because he stopped to watch a pod of dolphins. As he hurried on afterwards he reckoned he had the perfect title for a novel: *Delayed by Dolphins*.

> The moral of my father's dolphin moment is that the wildlifer is always looking: the wild side of the brain is never off-duty. That brings unexpected rewards in all kinds of unexpected places.

Thursday 23rd

Two old friends, both great wildlifers and conservationists, had dropped by for lunch and a stroll around the marsh: a good occasion. Now I was sitting out on a breezy, sunny day, comfortable and quiet: the rattle of magpies, calls from a pheasant, Distant Ducks.

A harrier crossed the marshes with something of a swagger: a juvenile, giving a reassuring sense of continuity. A group of jackdaws called among themselves, then a little owl yelped and instantly got a reply. The two exchanged views for a few moments: bird communicating with bird, with me the privileged eavesdropper.

I picked up the harrier again, a little more distant, turning into the wind and making rather a mess of it: an ungainly wobble, instantly corrected. Low marks from the judge under the ash tree ... but the bird was building up those practice hours, creating muscle memory, becoming a fully qualified creature of the air.

A butterfly narrowly missed me: for once breaking the unities I stood and pursued it for a few paces until it revealed itself as a spotted wood, a rare example of a butterfly that has gone in for subtle colouration.

> In general a butterfly wants two things from its colours: (1) to be very easily seen and (2) to be very hard to see. This contradiction works because the bright colours tend to be on the upper wings, working as signals to the opposite sex, while the underneath is cryptic or plain black, so with folded wings many butterflies are surprisingly hard to see. But there's a great deal more to it than that: check out Philip Howse's quite remarkable *Seeing Butterflies*.

Friday 24th

It was sunny and breezy with an emperor dragonfly posing in the air above the dyke. A succession of pigeons went past: I raised the bins to the fourth and it was a kestrel. Perhaps the other three had all been falcons as well, falcons of fabulous rarity: gyr falcons, red-footed falcons, Eleonora's falcons, Amur falcons ... and perhaps not. A volley of gunfire somewhere behind me.

A squirrel bounced out of cover on the ground about six feet in front of me and stopped dead, looking at me. He was puzzled, for I was already stopped dead myself. We examined each other for a while: eventually the squirrel vanished back into the bushes, followed by his tail.

I could hear the warm, busy cawing of rooks, looked up and found fifty of them overhead, spiralling and soaring, disturbed by the shooting, using a gentle thermal to gain a little free lift so they could express their sense of disturbance without expending too much energy.

Most corvid flight is direct and purposeful, for they tend to travel from one place to another as the crow flies: this movement was glidey and unassertive, not committed to any one direction. They provided the answer to the murmuratious birds I had seen a few days ago: corvids just above the horizon, shifting and turning as they made their restless way to bed, or to roost, rather.

A young moorhen called from the dyke with the voice of a tortured kitten. For a while I tried to catch flying dragonflies in the bins, once or twice succeeding. It was good sport, bringing occasional vistas of beauty. Not very meditative, though.

> Monty Roberts, the great horse-whisperer, advises you never to 'snap your eyes' when dealing with a horse, always change the direction of your gaze by moving – 'sliding' – your eyes slowly. It works with other species: I was doing well as a squirrel-whisperer.

Sunday 26th

My year of sitting was about to reach completion, for this was the last day of the year that I had allotted for this journey. My diary told me I was in Africa once more, so as I took my seat on a cool, blustery day I allowed myself the luxury of thinking about the place. About now I should be travelling to Mfuwe airport in the Luangwa Valley to meet the clients, hoping they were all going to be nice, and then driving with them to the lovely Tafika Lodge on the banks of the Luangwa

River. We would arrive in darkness and dine together in the open. The following morning as the sun rose they would get up, walk a few yards to the edge of the river and gasp. What was this place? Was it real? Surely it was too beautiful to exist in this tough old world – perhaps the rescuers were even now digging their bodies from the wrecked plane while their souls were arriving in heaven ...

But I was gazing across another floodplain with another river before me. The swifts, long gone from here, would be flying over the Luangwa River any day now. I had no visible clients to show the wonders available here, but instead I have you, dear reader, and even then, as I made a note for you in my notebook, this was a thought that gave me a deep and subtle pleasure – for here was a wonder new-minted: a distant but utterly unmistakable marsh harrier, just this side of the river, a little below the treetops, making perfect the entire landscape as it travelled, balanced on the perfect dihedral of its wings: two perfect strokes of the Zen master's brush on the shifting air.

The tree above me was creaking hard as the gusts reached 30 mph; a green woodpecker (not a bee-eater) flew across in a series of peaks and troughs. And then the harrier was in sight again, a young bird savouring the gusts, a good wind for practice, for honing those skills. This was a great bird in the making: and it would be better still the following year.

- The Africa trips take our guests thrillingly (yet safely) close to lions and elephants, often as we travel through the bush on foot. But every time, when we talk on the last night about the best moment of the trip, our guests come back to some moment of perfect contemplative calm: sitting beneath a tree as the life of the Valley gets on with its ancient job of teeming.

ARRIVAL

Monday 27th

A year is a conclusion: a year and a day is continuity, rebirth, eternal reiteration. A squally morning had given way to an afternoon of wind with outbreaks of a sun that gave little warmth. Tomorrow I would wear an extra layer, for I would journey on. I would sit here again, again and again, thinking and not thinking, seeking and not seeking. New beginnings make endings bearable.

My first unbelieving shock at my father's death was modulating into a good strong sadness, something I will never be without or wish to be. I had shown him many birds and I had taught him how to listen to birdsong: what more could anybody do for anyone?

We had a bet on one week-long Cornish walk: first one to see a seal gets a bottle of Glenmorangie. It was the last day: we had paused for a sit at Stackhouse Cove and I was staring hard at the sea. 'No seals down there, boy!'

'Try looking there, left of the big rock.'

'You bastard!'

Well, he should know.

I felt another kind of loss as well: the sort of thing you always feel at the end of a project, a loss inevitably combined with the thrill of the next: a new book, yes, and quite separately, a new scheme for observing and recording the life of the landscape all around me.

A black-headed gull flew one way, a crow the other: and as they passed in mid-air I wondered if they were providing me with the perfect downbeat ending: a gentle fade-out that that carries with it the acceptance that it is the ordinary things that had made the entire year so marvellous.

Then once more the harrier, a youngster relishing the wind. Riding the gusts the bird rose, fell back, rose again, unflapping, unflappable, lit in a single shaft of sunlight. Beneath my seat the earth kept on orbiting.

> 🍃 And now, dear reader, may I invite you to find your own spot? Take a seat, for the world is yours.